EARTH SCIENCES IN THE 21ST CENTURY

VOLCANIC ASH

CHEMICAL COMPOSITION, ENVIRONMENTAL IMPACT AND HEALTH RISKS

EARTH SCIENCES IN THE 21ST CENTURY

Additional books in this series can be found on Nova's website under the Series tab.

Additional e-books in this series can be found on Nova's website under the e-book tab.

EARTH SCIENCES IN THE 21ST CENTURY

VOLCANIC ASH

CHEMICAL COMPOSITION, ENVIRONMENTAL IMPACT AND HEALTH RISKS

DANIELLE GRAVER
EDITOR

NOVINKA

New York

Library of Congress Cataloging-in-Publication Data

Volcanic ash : chemical composition, environmental impact, and health risks / Danielle Graver, editor.
    pages cm. -- (Earth sciences in the 21st century)
  Includes index.
  ISBN 978-1-63463-597-4 (softcover)
  1. Volcanic ash, tuff, etc. 2. Volcanic ash, tuff, etc.--Analysis. 3. Volcanic ash, tuff, etc.--Environmental aspects. I. Graver, Danielle, editor.
  QE461.V615 2015
  552'.23--dc23   2014044413

Published by Nova Science Publishers, Inc. † New York

CONTENTS

PREFACE

Ash produced as a consequence of explosive volcanic eruptions can cause multiple hazards both close to the volcano and at great distances. Explosive volcano eruptions often release volcanic plumes into the atmosphere, which consist of tephra (submillimeter-sized rock particles), water vapor and other gases such as carbon dioxide (CO_2), sulfur dioxide (SO_2) and hydrogen sulfide (H_2S). Particles from volcano eruptions are transported by wind to thousands of kilometers away, or even over 10,000 km from their source for some fine particles. This book discusses the environmental impact and health risks volcanic ash poses as well as its chemical composition.

Chapter 1 – Volcanic ash clouds are a natural hazard with broad environmental impact. In this study, a new method is developed to predict the transport of volcanic plumes in the atmosphere. In contrast to many existing volcanic ash transport and dispersion (VATD) models that simulate the evolution of volcanic plumes, the new method focuses on the overall properties of the wind field in which volcanic particles are transported and uses a dynamics systems analysis to identify the attracting structures that dictate the transport. As demonstrated in the study of the eruptions of Augustine volcano in January 2006, these structures act as attractors in the atmosphere towards which volcanic ash particles are transported. These attracting structures are associated with hazard zones with high concentrations of volcanic ash. The advantages of the method are that the attracting structures are independent of particle source parameters and are less prone to inaccuracy in the wind field than particle trajectories. The new approach provides the hazard maps of volcanic ash, and is able to help improve long-term predictions and to estimate the environmental impact.

Chapter 2 – Natural disasters are able to cause significant damage to chemical plants, structure, infrastructure and lifelines (power plants, fuel storage facilities, water treatment plants, etc.). Failures in these systems could worsen the effects on people, environment and economy due to natural disasters. In the literature such scenarios are known as Natural-Technological (Na-Tech) events; in this frame volcanic eruptions can trigger a wide range of Na-Tech hazards for both human health and environment. This chapter extends a recent study related to the analysis of the potential damage of primary (mechanical) treatments included in wastewater plants due to the fallout of volcanic ash produced by volcanic explosive eruptions. The work aims at the identification of the conditions leading to the functionality reduction of grit removal facilities. An application of the proposed approach to a case study, which is the surrounding Mt. Etna (Italy), is also given.

Chapter 3 – Ash produced as consequence of explosive volcanic eruptions can cause multiple hazards both close to the volcano and at great distances. Ash fallout can produce a wide range of impacts on exposed assets (edifices, roads, lifelines), while airborne ash jeopardizes air travel safety. The attention on the impacts of ash dispersal on air traffic increased after the severe consequences of the 2010 Eyjafjallajökull eruption (Iceland), which caused an unprecedented closure of the European North Atlantic airspace. Thus, the need to predict the dispersal of volcanic ash became more important for both civil-defense authorities and airline stakeholders (companies, service providers). For this reason, tephra transport and dispersal models, which allow predicting tephra sedimentation and dispersal caused by a modelled eruptive scenario, became crucial in volcanic hazard and risk assessment. These models have been used for multiple purposes: short-term operational forecast of volcanic ash dispersion; reconstruction of past eruptions in order to constrain key eruption parameters; and generation of probabilistic hazard maps for hazard assessment of tephra dispersion and fallout. This chapter reviews the applications to long-term hazard assessment of volcanic ash dispersal and deposition for multiple eruptive scenarios at some of the most active volcanoes in the world. The authors present the improvements achieved during the last two decades focusing on recent developments of tephra dispersal hazard assessment, and we discuss some relevant aspects that must be improved in the future in order to optimize long-term hazard assessment strategies.

In: Volcanic Ash   ISBN: 978-1-63463-597-4
Editor: Danielle Graver   © 2015 Nova Science Publishers, Inc.

Chapter 1

ATTRACTING STRUCTURES IN VOLCANIC ASH TRANSPORT AND THE CORRELATION TO ENVIRONMENTAL HAZARD ZONES

Jifeng Peng[1]* *and Peter Webley*[2]

[1]Department of Mechanical Engineering, University of Alaska,
Anchorage, AK, US
[2]Geophysical Institute, University of Alaska, Fairbanks,
Fairbanks, AK, US

ABSTRACT

Volcanic ash clouds are a natural hazard with broad environmental impact. In this study, a new method is developed to predict the transport of volcanic plumes in the atmosphere. In contrast to many existing volcanic ash transport and dispersion (VATD) models that simulate the evolution of volcanic plumes, the new method focuses on the overall properties of the wind field in which volcanic particles are transported and uses a dynamics systems analysis to identify the attracting structures that dictate the transport. As demonstrated in the study of the eruptions of Augustine volcano in January 2006, these structures act as attractors in the atmosphere towards which volcanic ash particles are transported. These attracting structures are associated with hazard zones with high concentrations of volcanic ash. The advantages of the method are that the

* Correspondence to: jpeng@alaska.edu

attracting structures are independent of particle source parameters and are less prone to inaccuracy in the wind field than particle trajectories. The new approach provides the hazard maps of volcanic ash, and is able to help improve long-term predictions and to estimate the environmental impact.

INTRODUCTION

Explosive volcano eruptions often release volcanic plumes into the atmosphere, which consist of tephra (submillimeter-sized rock particles), water vapor and other gases such as carbon dioxide (CO_2), sulfur dioxide (SO_2), hydrogen sulfide (H_2S), etc. Particles from volcano eruptions are transported by wind to thousands of kilometers away, or even over 10,000 km from their source for some fine particles (Ram and Gayley, 1991). Volcanic plumes can reach over 20 km in altitude above sea level (Holasek et al., 1996), well into and above the cruising altitude of commercial aircrafts, thus they pose significant threats to aviation safety. A recent analysis of aircraft encounters with volcanic plumes indicates that there were 129 probable encounters between 1953 and 2009 (Guffanti et al., 2010). Furthermore, 79 of these encounters reported various degrees of damage to either airframes or engines, though no crash was caused. An encounter with volcanic ash can be costly, such as the 2000 encounter with a cloud from Miyakejima that resulted in $12 million of damage to a single aircraft (Tupper et al., 2004). In the Eyjafjallajökull eruption in Iceland during April and May 2010, ash clouds from the eruption reached thousands of kilometers away from volcano (Gudmundsson et al., 2010) and resulted in a week-long closure of a vast airspace in western Europe, during which over 100,000 flights were canceled, causing an unprecedented disruption in air travel and a loss of billions of dollars revenue for airlines (Wall and Flottau, 2010). Near ground surface ash and its deposits can be dangerous as well. Exposure to SO_2 and H_2S can cause many health problems including eye irritation, bronchitis, and sore throat (Delmelle et al., 2002). When inhaled, ash particles adversely affect the lungs, causing respiratory problems such as asthma, bronchitis, and silicosis (Horwell and Baxter, 2006). Although volcanic ash sometime has a beneficial long-term effect on agricultural soil, deposits have a short-term deleterious effect on crops (Self, 2006).

Over the past decades, many volcanic ash transport and dispersion (VATD) models have been developed for use by Volcanic Ash Advisory

Centers (VAACs) and other response centers, with examples including MEDIA (Piedelievre et al., 1990), HYSPLIT (Draxler and Hess, 1998), PUFF (Searcy et al., 1998), CANERM (D'Amours et al., 1998), and NAME (Jones et al., 2007). Coupled with Numerical Weather Prediction (NWP) data, these models are able to predict volcanic ash transport and dispersion in the atmosphere, providing scientific basis for authorities to plan hazard responses.

Although these VATD models are remarkably useful, there are still many limitations on wider applications of these models. The biggest challenge is the prediction accuracy, which depends mainly on two factors. First, many VATD models use Lagrangian particle trajectories, whose accuracy is highly dependent on the quality of NWP data, especially its spatial and temporal resolution. Compared with reference trajectories, the mean error for ash particle trajectories over 36h travel time can be as high as 35% (Stunder, 1996). Using finer-grid NWP data would improve prediction, however, high-resolution NWP simulations generally require considerable amount of time. The delay in the availability of high quality NWP data deters fast and accurate ash transport prediction.

Another factor that restricts wider applications on real time monitoring and response is the lack of accurate descriptions of the initial eruptive plume. The initial parameters are critical to the success of many VATD models, because even short term predictions heavily depend on descriptions of initial plume, e.g. maximal height, particle size and number density distribution, etc. However, it is usually difficult to collect these data, especially during the early hours after the eruption, when a fast response is desired. Eruptive volumes normally have distinct flow characteristics at various heights and therefore comprise distinct layers with different ash properties (Kieffer, 1984). Moreover, values of these source parameters may change during eruption (Mastin et al., 2009). Coordinated and multidisciplinary efforts have been used to improve the accuracy of source parameters, by establishing correlation between these parameters from well-documented past eruption events (Mastin et al., 2009), but there is still a level of uncertainty in the defined input parameters (Webley and Steensen, 2013).

Therefore, the utilization of conventional VATD models is limited by the accuracy of both the physical properties of initial ash plume and the wind field in which particles are transported. To overcome these limitations, a new approach was proposed to provide a fast, accurate prediction on volcanic plume transport that is less dependent on the quality of NWP data and initial ash plume parameters. In contrast to the previous methods that simulate evolution of plumes, the new method focuses on the overall properties of the wind field in which volcanic particles are transported and correlates particle motion to some underlining structures that dictate the transport. To be more specific, the method uses a dynamical systems approach to identify attracting structures in the wind field. These structures act as global attractors to which particles move towards. These structures, as the underlining structure of wind transport process, are independent to particle source parameters. The method was demonstrated and compared with an existing VATD model in a study on the Eyjafjallajökull eruption in Iceland during April 2010 (Peng and Peterson, 2012). The study demonstrated these structures agree with high particle concentration of volcanic clouds from the VATD model. In this study of the eruptions of Augustine volcano in January 2006, the method of attracting structures is first compared with satellite images of volcanic clouds. The attracting structures coincide with volcanic plumes from satellite imagery, therefore indicating hazard zones from volcano plumes.

The 2006 Eruption of Augustine Volcano

Augustine is a historically active volcano on Augustine Island, Alaska. The 2006 eruption of Augustine consisted of four phases defined by the characteristics of eruptive activity. These phases are the precursory (May 2005 to January 2006), the explosive (January 11 to 28, 2006), the continuous (January 28 to February 2, 2006), and the effusive (February to late March, 2006). Monitoring instruments recorded several powerful explosions that occurred between 13:24 UTC on January 13 and 09:14 UTC on January 14, 2006. Plumes reached altitudes of 14 kilometers as they moved eastward and disrupted commercial airline traffic (Power et al., 2010).

Weather Forecast Data

The wind data from a Weather Research and Forecasting (WRF) model simulation during the Augustine volcano eruption in 2006 was used for this study. The WRF simulation was initialized using the global GFS simulation at 00:00 Coordinated Universal Time (UTC) January 13, 2006, and ran for 72 hours with an output at every 1 hour. The domain is a polar stereographic grid (299 × 199) with a spatial resolution of 18 km centered at the volcano (59.4°N 153.4°W). There are 34 vertical levels spanning 100 to 20000 meters in elevation. All the analysis in this study was based on this high resolution WRF data.

Finite Time Lyapunov Exponents and Attracting Structures

Instead of modeling volcanic ash transport based on the evolution of the plume from an eruption, a dynamical systems approach is used to identify global coherent structures in the wind field. We use the finite time Lyapunov exponents (FTLE) to locate attractors in a velocity field (Haller and Yuan, 2000; Haller, 2001; Shadden et al., 2005). FTLE describe the maximal rate of extension of a line element advected in the flow. In other words, FTLE measure the maximal separation/converging rate of nearby tracers in the flow.

Assuming that a tracer particle in the wind field follows the wind velocity, i.e., $\mathbf{v}(t;t_0,\mathbf{x}_0) = \mathbf{w}(t,\mathbf{x})$. Given its velocity $\mathbf{v}(t;\ t_0,\ \mathbf{x}_0)$ at a series of time instants, the Lagrangian trajectory of a tracer particle $\mathbf{x}(t;\ t_0,\ \mathbf{x}_0)$ is the solution of

$$d\mathbf{x}(t;t_0,\mathbf{x}_0)/dt = \mathbf{v}(t;t_0,\mathbf{x}_0), \qquad (1)$$

with the initial condition $\mathbf{x}(t_0;t_0,\mathbf{x}_0) = \mathbf{x}_0$. Notice that location and velocity of the particle is Lagrangian marked by $(t_0,\ \mathbf{x}_0)$ whereas the velocity of wind field is Eulerian.

To avoid confusion between particle trajectories and spatial coordinates, the notation ϕ is used for particle trajectories and $\mathbf{x}$ for spatial coordinates. By following particle trajectories over a duration of time T after initial time t_0, a flow map $\phi_{t_0}^T(\mathbf{x})$ is obtained that maps particles from their position $\mathbf{x}$ at initial time t_0 to their position at time $t = (t_0 + T)$, as

$$\phi_{t_0}^T(\mathbf{x}) = \mathbf{x}(t_0 + T;t_0,\mathbf{x}_0) = \phi_{t_0}^0(\mathbf{x}) + \int_{t_0}^{t_0+T} \mathbf{u}(\tau,\mathbf{x}(\tau;t_0,\mathbf{x}_0))d\tau. \qquad (2)$$

From the flow map $\phi_{t_0}^T(\mathbf{x})$, we compute the FTLE as

$$\sigma_{t_0}^T(\mathbf{x}_0) = \frac{1}{|T|} \ln \lambda_{max}(t;\mathbf{x}_0,t_0), \qquad (3)$$

where $\lambda_{max}(t;\ \mathbf{x}_0,\ t_0)$ is the square root of the largest eigenvalue of the right Cauchy-Green deformation tensor.

$$\Delta = \left[\frac{d\phi_{t_0}^T(\mathbf{x})}{d\mathbf{x}} \right]^* \cdot \left[\frac{d\phi_{t_0}^T(\mathbf{x})}{d\mathbf{x}} \right]. \tag{4}$$

In Eq. $\mathbf{D} = d\phi_{t_0}^T(\mathbf{x})\,/\,d\mathbf{x}$ is the deformation gradient tensor. The denotation '*' means the transpose of a tensor and '·' represents the tensor product.

The physical meaning of FTLE $\sigma_{t_0}^T(\mathbf{x}_0)$ can be explained by following particle trajectories over a duration of time T after initial time t_0. Consider the trajectories for a slightly perturbed particle at $\mathbf{y} = \mathbf{x} + \delta\mathbf{x}(0)$ at time t_0. After a time interval T, this perturbation becomes

$$\delta\mathbf{x}(T) = \phi_{t_0}^T(\mathbf{y}) - \phi_{t_0}^T(\mathbf{x}) = \frac{d\phi_{t_0}^T(\mathbf{x})}{d\mathbf{x}}\delta\mathbf{x}(0) + O\left(\|\delta\mathbf{x}(0)\|^2\right) = \mathbf{D}\delta\mathbf{x}(0) + O\left(\|\delta\mathbf{x}(0)\|^2\right). \tag{5}$$

By dropping high order terms of $\delta\mathbf{x}(0)$, the magnitude of the perturbation is given by

$$\|\delta\mathbf{x}(T)\| = \sqrt{\delta\mathbf{x}(0) \cdot \Delta \cdot \delta\mathbf{x}(0)}\ .$$

The magnitude of the perturbation is maximal when $\delta\mathbf{x}(0)$ is aligned with the eigenvector associated with the maximum eigenvalue of Δ. That is, if $\lambda_{max}(\Delta)$ is the square root of the maximum eigenvalue of Δ, then

$$\|\delta\mathbf{x}(T)\| = \lambda_{max}(\Delta)\|\delta\mathbf{x}(0)\|.$$

$$\tag{6}$$

Therefore the FTLE represents the maximum linear growth rate of a small perturbation,

$$\sigma_{t_0}^T(\mathbf{x}) = \frac{1}{|T|} \ln \lambda_{\max}(\Delta) = \frac{1}{|T|} \ln \frac{\|\delta\mathbf{x}(T)\|}{\|\delta\mathbf{x}(0)\|}. \tag{7}$$

The analysis above can be applied to two-dimensional (2D) as well as three-dimensional (3D) velocity fields. FTLE fields are scalar fields and visualized as contour plots. The ridges on FTLE contour plots, which have local maximal values, indicate the structures that have the maximal separation and converging rate. Intuitively, for a 2D FTLE, a ridge line is a curve normal to which the topography is a local maximum. Similarly, for a 3D FTLE, a ridge is a surface normal to which the topography is a local maximum. The ridges, i.e., level sets of maximal separation and converging rate, represent either repelling structures or attracting structures. When fluid particle trajectories are integrated forward in time (i.e. T > 0), ridges of the forward FTLE are repelling structures. These structures are said to be repelling because particles on either side of the structures are strongly repelled. Conversely, backward-time integration of fluid particle trajectories (T < 0) calculate backward FTLE whose ridges reveal attracting structures, along which fluid particles on either side of the structures are attracted to them. Only the attracting structures are used in this study because they are of interests to this study.

COMPUTATION OF FTLE AND IDENTIFICATION OF ATTRACTING STRUCTURES

The input, the wind velocity field from NWP data, consists of a time sequence of velocity data defined on a mesh of discrete points. To calculate FTLE $\sigma_{t_0}^T(\mathbf{x})$, a region of interest is first defined on which FTLE will be calculated. A Cartesian mesh is constructed over the region and used as the initial grid. For every point on the initial grid, its trajectories from $\mathbf{x}(t_0)$ to $\mathbf{x}(t_0 + T)$ is calculated. A 4-th order Runge-Kutta algorithm is used to calculate the integration. A cubic interpolation scheme is used to compute the velocity at arbitrary positions (Lekien and Marsden, 2005).

After trajectories are calculated, the (right) Cauchy-Green deformation tensor $d\phi_{t_0}^T(\mathbf{x})/d\mathbf{x}$ is evaluated for every node on the initial mesh. For a 2D velocity field, the deformation tensor at a given node $\mathbf{x} = (x_{i,j}, y_{i,j})$ is given by

$$\frac{d\phi_{t_0}^T(\mathbf{x})}{d\mathbf{x}}\bigg|_{(x_{i,j}(t),y_{i,j}(t))} = \begin{bmatrix} \dfrac{x_{i+1,j}(t_0+T)-x_{i-1,j}(t_0+T)}{x_{i+1,j}(t_0)-x_{i-1,j}(t_0)} & \dfrac{x_{i,j+1}(t_0+T)-x_{i,j-1}(t_0+T)}{y_{i,j+1}(t_0)-y_{i,j-1}(t_0)} \\[4mm] \dfrac{y_{i+1,j}(t_0+T)-y_{i-1,j}(t_0+T)}{x_{i+1,j}(t_0)-x_{i-1,j}(t_0)} & \dfrac{y_{i,j+1}(t_0+T)-y_{i,j-1}(t_0+T)}{y_{i,j+1}(t_0)-y_{i,j-1}(t_0)} \end{bmatrix}. \tag{8}$$

The largest eigenvalue of Δ is calculated to determine FTLE $\sigma_{t_0}^T(\mathbf{x})$ at every node. This process can be repeated to calculate FTLE for different t_0 at a series of time frames. Interested readers can refer to Leiken et al. (2007) for 3D FTLE calculation.

The extraction of ridges from FTLE fields can be accomplished by a variety of ad hoc methods including thresholding or gradient searches of the FTLE field to identify local maxima. In this study, the ridges are extracted visually from FTLE contour plots.

COMPARISON WITH SATELLITE IMAGES

To demonstrate the utilities of the attracting structures in volcanic plume transport, these structures were compared with satellite images of volcanic plumes. The volcanic plumes were visualized by the Advanced Very High Resolution Radiometer (AVHRR) sensor on NOAA satellites. The AVHRR uses the temperature difference in two long wave IR bands, namely at 11 and 12 μm, to effectively observe thin plumes (Prata 1989). The AVHRR data were collected by receiving stations operated by the Geographic Information Network of Alaska at the University of Alaska's Geophysical Institute. For details of the methodology used in the volcanic ash detection, please refer to Power (2010).

RESULTS AND DISCUSSIONS

The 2D wind field from the WRF data at the altitude of 10 km was used to determine the attracting structures. A snapshot of the FTLE field of this 2D wind field is plotted in Figure 1a. Multiple attracting structures are extracted from the FTLE contour and are plotted in Figure 1b together with the velocity vector field of that time instant. It is clear that the one can not determine the structures directly from the Eulerian velocity field. In other words, the pattern of instantaneous velocity field does not correlate directly with the geometry of attracting structures.

The attracting structures represent attractors in the flow fluid. Because the flow field is time-dependent, the attracting structures also evolve with time. To illustrate that volcanic clouds are attracted to these structures over time, we superimpose the attracting structures on the AVHRR satellite images. Figure 2 shows the attracting structures on the AVHRR satellite images of volcanic plumes from three eruptions of Augustine volcano in the morning of January 13, 2006.

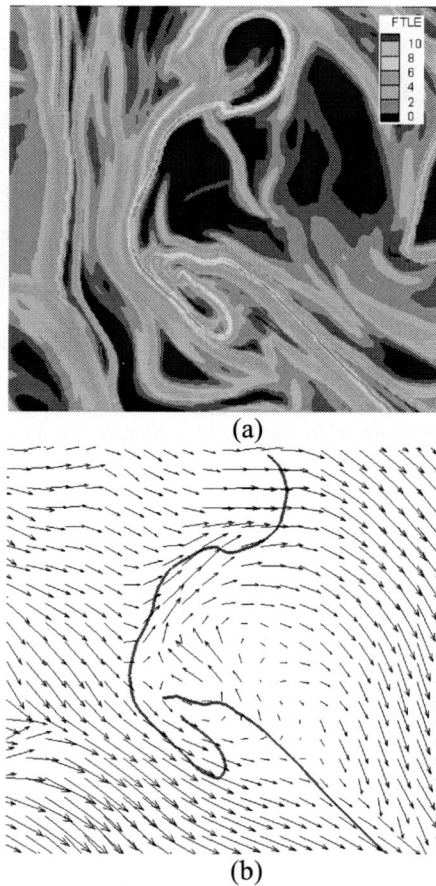

(a)

(b)

Figure 1. (a) The 2D FTLE field at 10 km altitude above sea level. (b) the attracting structures (red), which are the level set of largest values of FTLE, superimposed with wind velocity vector field.

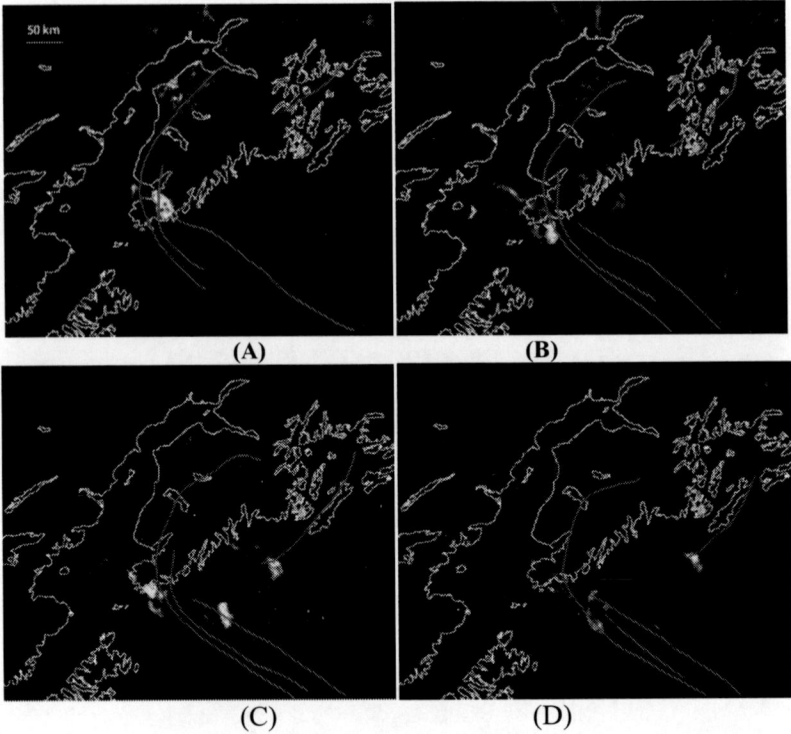

Figure 2. Temporal evolution of the attracting structures (red curves) and the motion of an ash plume (contours) shown on band 4 minus band 5 AVHRR images from the eruptions of Augustine volcano on January 13, 2006. The 4 different frames are at (a) 17:34, (b) 20:39, (c) 22:21 and (d) 23:27 UTC. The evolution of volcanic plumes follows that of the attracting structures.

The 4 different frames are at (a) 17:34, (b) 20:39, (c) 22:21 and (d) 23:27 UTC, January 13, 2006. It is clearly seen that most of the volcanic plumes are located at or in very close proximity of the attracting structures. The attracting structures not only coincide with high-concentration volcanic plumes, but also the low-concentration ones. From the evolution of attracting structures and volcanic plumes, it is also shown that the volcanic clouds follow the attracting structures. Therefore, the attracting structures correlate strongly with the location of volcanic plumes and can be used to indicate hazard zones from volcano eruptions.

The plots in Figure 2 show the evolution of volcanic plumes within 24 hours of eruptions and a small distance (less than 500 km) from the volcano. Satellite images of volcanic plumes long after eruptions are less available because the plumes usually become very thin over time and difficult to detect. Though most large, heavy particles settle to the ground in hours, fine particles can keep floating in the atmosphere for days or longer and travel over thousands of kilometers away from the source (Ram and Gayley, 1991). These fine particles pose a significant risk to the environment and aviation safety and thus requires long-term monitoring approaches other than satellite imagery. A previous study (Peng and Peterson, 2012), in which the attracting structures were compared with a VATD model simulation of volcanic plume transport, has demonstrated that in the long term, the method of attracting structures is more reliable than particle trajectories, which make them more valuable for extended predictions of plumes with high accuracy. The ability of long-term prediction with high accuracy can be particularly useful to identify hazard zones after a volcano eruption has elapsed. This would enable authorities to estimate and response to the environmental effects of volcanic ash plumes.

Limited by the information by the satellite images, only 2D analysis was used and the attracting structures at a given altitude was determined and compared with the volcanic plumes images. However, the method of attracting structures can also be applied to 3D space (Peng and Peterson, 2012). Instead of being 2D curves on a plane, the attracting structures are surfaces in 3D space. Even though ash particles gradually fall in altitude, they are still attracted to the attracting structures. The reason is that wind transport is the attracting mechanism for particles to move towards these structures. In essence, there is competition between wind transport, which move particles towards the attacking structures, and gravitational settling, which move particles away from them. So if there were no gravitation, particles would strictly move towards the 3D attracting structures and stay on them. On the other hand, gravitational settling can be interpreted as a disturbance to the wind transportation and it makes particles move away (vertically) from the 3D attracting structure. So considering that a particle falls out from the 3D attracting structure at an altitude due to gravitation, the wind will keep carrying it back towards the attracting structures at a new altitude.

The attracting structures based on FTLE are more coherent than arbitrary structures and are robust to even large errors of the velocity field as long as these errors are local in time (Haller, 2002). The errors in calculating attracting structures are much less than those of particle trajectories, thus enable the analysis to generate reliable results from even approximate, low resolution NWP data. The study by Peng and Peterson (2012) demonstrated that when calculated from a low-resolution NWP data, attracting structures can achieve higher accuracy compared with Lagrangian trajectories, thus make this new method less prone to errors in NWP data.

The computational cost of FTLE and the attracting structures is sufficiently low to guarantee fast analysis and response. Using a single processer personal computer, the 2D analysis takes several minutes and a 3D analysis usually less than an hour. The computation can speed up using high performance computers. Because NWP models usually need longer computational time, apparently the availability of high-resolution NWP data is the limit in time consumption of the analysis, not the FTLE computation. In addition, to prevent large errors in particle trajectories, existing VATDs constantly use updated NWP data to recalculate. However, because the attracting structures are less prone to errors in NWP data than particle trajectories, the new method does not require as frequent recalculations, with the potential to overcome the limitation on the availability of frequent update of high-resolution NWP data.

The focus of future studies is to implement the method in real-time monitoring and to assist authorities in response to volcanic clouds. The method would be used as complement to remote sensing and detection of volcanic plumes from satellites. Satellite images are able to provide accurate information on the location of volcanic ash plumes. The analysis of plume position based on satellite images can be constantly input to the prediction model to further improve accuracy.

REFERENCES

D'Amours R., Servranckx R., Toviessi J.P., Trudel S., 1998. The operational use of the Canadian Emergency Response Model: Data assimilation, processing, storage, and dissemination. In: OECD Nuclear Energy Agency (Editor), Nuclear Emergency Data management: Proceedings of an International Workshop. Organisation for Economic Cooperation and Development. *Nuclear Energy Agency*, 215-221.

Delmelle P., Stix J., Baxter P., Garcia-Alvarez J., Barquero J., 2002. Atmospheric dispersion, environmental effects and potential health hazard associated with the low-altitude gas plume of Masaya volcano, *Nicaragua*. *Bulletin of Volcanology* 64, 423–434.

Draxler R.R., Hess G.D.,1998. An overview of the Hysplit 4 modeling system for trajectories, dispersion, and deposition. *Australian Meteorological Magazine* 47, 295-308.

Guffanti M., Casadevall T.J., Budding K., 2010. Encounters of aircraft with volcanic ash clouds; A compilation of known incidents, 1953–2009. *U.S. Geological Survey Data Series* 545.

Gudmundsson M.T., Pedersen R., Vogfjör K., Thorbjarnardóttir B., Jakobsdóttir S., Roberts M.J. 2010. Eruptions of Eyjafjallajökull Volcano, Iceland. *Eos, Transactions American Geophysical Union*, 91, 190-191.Haller G., 2002. Lagrangian coherent structures from approximate velocity data. *Physics of Fluids* 14, 1851-1861.

Haller G., 2001. Distinguished material surfaces and coherent structures in 3D fluid flows. *Physica D* 149, 248-277.

Haller G., Yuan G., 2000. Lagrangian coherent structures and mixing in two-dimensional turbulence. *Physica D* 147, 352-370.

Holasek R.E., Self S., Woods A.W., 1996. Satellite observations and interpretation of the 1991 Mount Pinatubo eruption plumes. *Journal of Geophysical Research* 101, 27635–27655.

Horwell C.J., Baxter P.J., 2006. The respiratory health hazards of volcanic ash: a review for volcanic risk mitigation. *Bulletin of Volcanology* 69, 1–24.

Jones A., Thomson D., Hort M., Devenish B., 2007. The U.K. Met office's next-generation atmospheric dispersion model, NAME III. In: Borrego, C., Norman, A.L. (Eds.), Air Pollution Modeling and Its Application XVII. Springer, Berlin, 580-589.

Kieffer S., 1984. Factors governing the structure of volcanic jets, Explosive Volcanism: Inception, Evolution and Hazards, National Academy Press, Washington DC, 143-157.

Lekien F., Marsden J.E., 2005. Tricubic interpolation in three dimensions *Int. J. Numer. Meth. Engng* 63, 455–471.

Mastin L.G., Guffanti M., Servranckx, R., etc, 2009. A multidisciplinary effort to assign realistic source parameters to models of volcanic ash-cloud transport and dispersion during eruptions. *Journal of Volcanology and Geothermal Research* 186, 10-21.

Peng J., Peterson, R. 2012. Attracting structures in volcanic ash transport. *Atmospheric Environment* 48, 230-239.

Piedelievre J.P., Musson-Genon L., Bompay F., 1990. MEDIA — An Eulerian model of atmospheric dispersion: First validation on the Chernobyl release. *Journal of Applied Meteorology* 29, 1205-1220.

Prata F., 1989. Observations of volcanic ash clouds in the 10-12 mm window using AVHRR/2 data. *International Journal of Remote Sensing*, 10, 751-761.

Power J.A., Coombs M.L., Freymueller J.T., 2010. The 2006 Eruption of Augustine Volcano, Alaska. U.S. Geological Survey Professional Paper 1769.

Ram M., Gayley R.I., 1991. Long-range transport of volcanic ash to the Greenland ice sheet. *Nature* 349, 401-404.

Searcy C., Dean K., Stringer W., 1998. PUFF: A high-resolution volcanic ash tracking model. *Journal of Volcanology and Geothermal Research* 80, 1-16.

Self S., 2006. The effects and consequences of very large explosive volcanic eruptions. *Phil. Trans. R. Soc. A* 364, 2073-2097.

Stunder B.J.B., 1996. An Assessment of the Quality of Forecast Trajectories. *Journal of Applied Meteorology* 35, 1319-1331.

Tupper A., Kamada Y., Todo N., Miller E., 2004. Aircraft encounters from the 18 August 2000 eruption at Miyakejima, Japan. *Proceedings of the 2nd International Conference on Volcanic Ash and Aviation Safety,* Alexandria, VA 21–24 June 2004. National Oceanic and Atmospheric Administration, Silver Springs, MD USA, 5–9.

Wall R., Flottau J., 2010. Out of the ashes: Rising losses and recriminations rile Europe's air transport sector. *Aviation Week & Space Technology* 172, 23-25.

Webley P.W., Steensen T. 2013. Operational Volcanic Ash Cloud Modeling: Discussion on Model Inputs, Products, and the Application of Real-Time Probabilistic Forecasting. Lagrangian Modeling of the Atmosphere, Wiley, New Jersey, 271-298.

In: Volcanic Ash   ISBN: 978-1-63463-597-4
Editor: Danielle Graver   © 2015 Nova Science Publishers, Inc.

Chapter 2

THE IMPACT OF VOLCANIC ASH FALLOUT ON INDUSTRIAL FACILITIES: NATURAL-TECHNOLOGICAL HAZARDS IN WASTEWATER TREATMENTS (GRIT REMOVALS)

G. Ancione[1], P. Primerano[1], E. Salzano[2], G. Maschio[3] and M. F. Milazzo[1] *

[1]Dipartimento di Ingegneria Elettronica, Chimica e Ingegneria Industriale, University of Messina, Messina, Italy
[2]Istituto Ricerche sulla Combustione, Consiglio Nazionale delle Ricerche, Napoli, Italy
[3]Dipartimento di Ingegneria Industriale, University of Padova, Padova, Italy

ABSTRACT

Natural disasters are able to cause significant damage to chemical plants, structure, infrastructure and lifelines (power plants, fuel storage facilities, water treatment plants, etc.). Failures in these systems could worsen the effects on people, environment and economy due to natural disasters. In the literature such scenarios are known as Natural-

---
* Corresponding author: mfmilazzo@unime.it

Technological (Na-Tech) events; in this frame volcanic eruptions can trigger a wide range of Na-Tech hazards for both human health and environment. This chapter extends a recent study related to the analysis of the potential damage of primary (mechanical) treatments included in wastewater plants due to the fallout of volcanic ash produced by volcanic explosive eruptions. The work aims at the identification of the conditions leading to the functionality reduction of grit removal facilities. An application of the proposed approach to a case study, which is the surrounding Mt. Etna (Italy), is also given.

1. INTRODUCTION

Natural technological events (or simply Na-Techs) are technological accidents triggered by natural phenomena. Na-Techs increased over these last decades due to the increasing industrialisation of several territories and, in some cases, also to climate changes; Fabiano and Currò showed that nearly 6.4% of the accidents in the oil industry are due to natural phenomena [1]. Nevertheless, in several contexts, it has been highlighted that many European countries have to face various natural hazards, but appropriate management plans, based on the definition of the hazards associated with technological accidents triggered by natural events, are not designed [6]. Thus worldwide scientific community has recently pointed the need to define extended approaches for the risk analysis [2-5] to support the management of Na-Techs.

Amongst many natural phenomena, a growing attention has recently been paid to ash emissions as they may be a significant hazard for aviation. The most critical effect on aircrafts is caused by the ash melting in the hot section of the engine and, then, fusing into a glass-like coating on components further back in the engine. As a consequence a loss of thrust and a possible engine failure could be determined; in addition an abrasion of engine parts and a possible clogging of the fuel and the cooling systems occur [7]. The effects of volcanic ash fallouts on chemical installations and lifelines (electrical power grids, water distribution systems and gas and oil pipelines) could also be significant. Several examples of Na-Techs caused by ash emissions are given by the literature, some of them occurred during the eruption of St. Helen in 1980; this event had a noteworthy impact on wastewater treatment plants [8]. Until now the impact of ash fallout has been studied with respect to the following targets: water treatment systems [9-11], transportation of hazardous materials [11], buildings [12], electric motors [13], atmospheric storage tanks for flammable liquids [14, 15] and filtering systems [16].

This chapter focuses on the study of the effects of volcanic ash on grit removals, which are included in mechanical (or primary) treatment units of Wastewater Treatment (WWT) plants. Mechanical treatment units are the most vulnerable to damages induced by ash fallouts, as these are the first ones encountered by the sewage to be purified. A previous study [9] led to the identification of all potential causes of failure in a generic WWT and, in particular, in facilities used for mechanical treatments; as shown in [10], the size of the particles emitted during explosive eruptions oriented towards the study of the effects on equipment engaged for screening and sedimentation processes. The present chapter aims to investigate the conditions leading to the incomplete removal of the particles in grit removal facilities. On the basis of the results, either equipment design proposals or alternatives solutions will be formulated accounting for the specificity of the territory where the WWT is located.

The chapter is structured as follows: Section 2 provides an overview on the solid removal in wastewater treatment plants; Section 3 gives a description of the approach for the estimation of the amount of removed materials, which allows estimating the functionality reduction of the equipment; Section 4 introduces the case-study which will be used to apply the approach proposed in Section 3; and, finally, Section 5 shows the results and gives a discussion about them.

2. SOLIDS REMOVAL IN WASTEWATER TREATMENTS

In a WWT plant the sewage is treated to enable it to be discharged into a watercourse or reused in a certain cycle. The wastewater is polluted by suspended solids, biodegradable organics, pathogenic bacteria and other disease and nutrients (including nitrates and phosphates) [17]. A wastewater treatment is essentially divided into three sections: *primary* (*mechanical*) *treatment*, *secondary* (*biological*) *treatment* and *tertiary treatment*, including several steps depending on the water quality requirements and the contamination of the sewage.

The focus of this chapter is on *primary* or *mechanical treatment*, in which suspended solids are mechanically removed from the sewage. Initially, materials such as pieces of wood, plastic bags and fabric are filtered out by using a bar screen, sometimes also a fine screen is included to remove small particles. Then the water flows into a grit chamber; in this sedimentation tank, mineral solids such as sand and gravel are separated by sedimentation.

Table 1. Main advantages and disadvantages of grit removal systems [19]

Apparatus	Advantages	Disadvantages
Aerated Grit Chamber	Consistent removal efficiency over a wide flow range. Low content of putrescible organic material with a well controlled aeration rate. Reduction of septic conditions in incoming wastewater by using pre-aeration. Versatility (allowing for chemical addition, mixing, pre-aeration and flocculation).	Potentially harmful volatile organics and odours may be released. More power than other grit removal processes. Additional maintenance and control of the aeration system than other grit removal processes.
Vortex-Type Grit Chamber	Removal of a high percentage of fine grit (up to 73 % of 0.11 mm size). Consistent removal efficiency over a wide flow range. No submerged bearings or parts that require maintenance. Smaller horizontal dimension compared to other systems (advantageous when space is an issue). Minimal head loss typically 6 mm.	Proprietary design, which makes modifications difficult. Paddles tend to collect rags. Deep excavation due to their depth, increasing construction costs, especially if rock is present. The grit sump tends to clog and requires high-pressure agitation using water or air to loosen grit compacted in the sump.
Horizontal Flow Grit Chamber	Flexible (allowing performance to be altered by adjusting the outlet flow control device). Easy construction.	Difficulty to maintain a 0.3 m/s velocity over a wide range of flows. Excessive wear to the submerged chain, flight equipment and bearings. Excessive head loss (typically 30 to 40 % of flow depth). Bottom scour due to high velocities at the channel bottom, with the use of weirs.
Hydrocyclone	Removal of both grit and suspended solids.	High energy requirement because of the use of pumps. Coarse screening is previous required to remove sticks, rags, and plastics.

Organic solids have a much lower settling velocity than sand and, consequently, a low sedimentation step is required to separate them. This process is named *primary clarification* and solids which separate at this stage are termed *primary sludge*.

A study of volcanic Na-Techs associated with ash fallouts in screening processes was faced by Milazzo and co-workers [10]; they investigated the effects on fine screens given that, as resulting from the particle size distribution analysis, these are able to retain such particles. Fine screens are not always included in wastewater treatments, thus the removal of particles is often provided by grit chambers. With respect to grit removal systems, *grit* is traditionally defined as particles larger than 0.21 mm and with a specific gravity (ratio of the material density to the density of water) greater than 2.65 [18].

Equipment design was traditionally based on removal of 95 % of these particles but, after the recent recognition that smaller particles must be removed to avoid damaging downstream processes, many modern systems are capable of removing up to 75 % of 0.15 mm material [19]. Grit removal facilities typically precede primary clarification and follow screening and comminuting/grinding (reducing the size of coarse solids). The main types of grit removal include aerated grit chambers, vortex-type grit removal systems, horizontal flow grit chambers (velocity-controlled channel) and cyclones (cyclonic inertial separation). Table 1 shows advantages and disadvantages of the grit removal systems which will be briefly described in the Section below [19].

2.1. Aerated Grit Chambers

In *aerated grit chambers*, grit is removed by causing the sewage to flow in a spiral pattern (Figure 1).

The air is introduced in the chamber along one side, this determines a perpendicular spiral velocity pattern through the tank. Heavier particles diverge from the streamlines and drop to the bottom of the tank, while lighter particles are eventually carried out of the tank.

This equipment is typically designed to remove particles of 0.21 mm or larger, with a detention period of two to five minutes. Proper adjustment of air velocity will result in nearly 100 % removal of the desired particle size [17, 19].

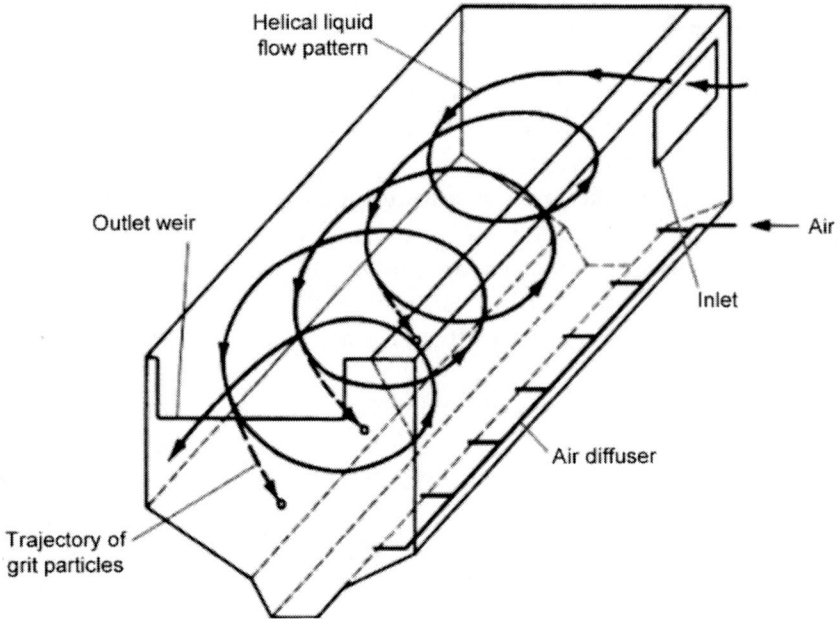

Figure 1. Aerated grit chambers (adapted from [17]).

2.2. Vortex Grit Chambers

The *vortex grit chambers* consist of a cylindrical tank in which the flow enters tangentially, creating a vortex flow pattern (Figure 2). Grit settles by gravity and may be removed by a pump.

Two chamber designs exist: (*i*) chambers with flat bottoms and a small opening to collect grit and (*ii*) chambers with a sloping bottom and a large opening into the grit hopper.

The sewage flow should be straight, smooth and streamlined. The ideal velocity range is typically 0.6 to 0.9 m/s. A minimum velocity of 0.15 m/s should be maintained at all times, because lower velocities will not carry grit into the grit chamber [17, 19].

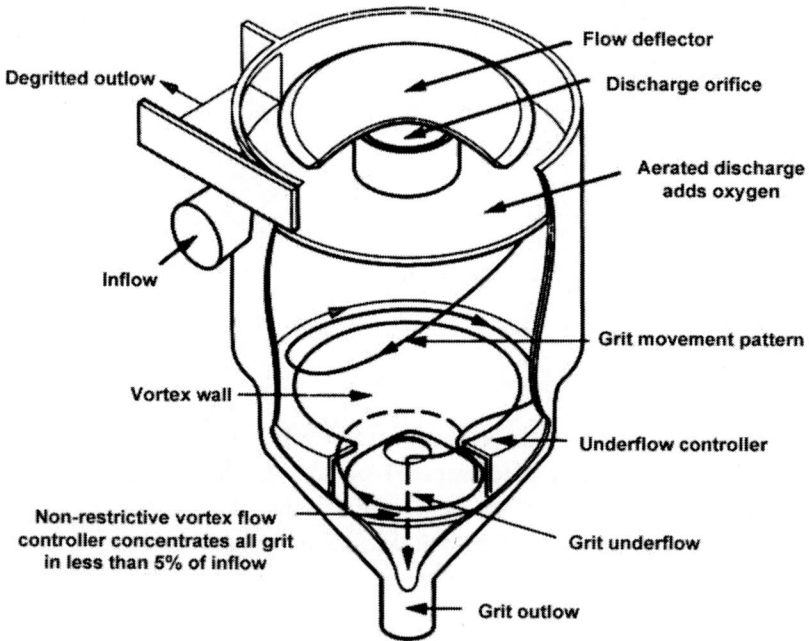

Figure 2. Vortex grit chambers (adapted from [17]).

2.3. Horizontal Flow Grit Chambers

The *horizontal flow grit chambers* (Figure 3) are the oldest type of grit removal system. Grit is removed by maintaining a constant upstream velocity of 0.3 m/s.

In these systems, heavier grit particles settle to the bottom of the channel, while lighter particles remain suspended and are transported out of the channel. Grit that does not require further classification may be removed with an effective flow control [17].

Proportional weirs or rectangular control sections are used to vary the depth of flow and keep the velocity of the flow stream at a constant 0.3 m/s. The length of the grit chamber is governed by the settling velocity of the target grit particles and the flow control section-depth relationship. An allowance for inlet and outlet turbulence is added.

The cross sectional area of the channel is determined by the rate of flow and the number of channels.

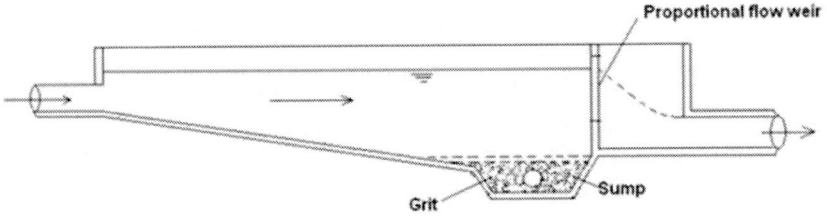

Figure 3. Horizontal flow grit chambers.

2.4. Hydrocyclone

In *cyclones*, heavier grit and suspended solids collect on the sides and bottom of the cyclone due to induced centrifugal forces, while scum and lighter solids are removed from the centre through the top of the cyclone. A cyclone can potentially remove as many solids as a primary clarifier.

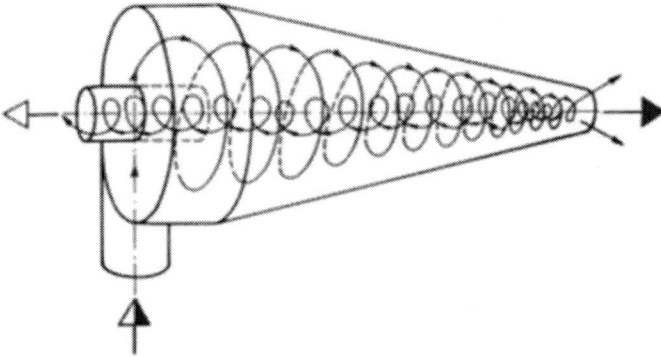

Figure 4. Hydrocyclone.

3. METHODOLOGY

The theoretical base of the solid removal from a liquid in grit chambers is the derivation of the *terminal settling velocity* (u_t). Assuming a low concentrated dispersion of particles, each particle settles discretely as if it is alone (unhindered by the presence of other particles). Starting from the rest, the settling velocity of the particle under gravity (u_s) in the fluid increases with the particle density; moreover it is accelerated until the resistance to the flow from the fluid equals the weight of the particle, then the settling velocity remains constant and this is named terminal settling velocity [20].

Given a fluid at constant temperature and a particle with constant settling velocity, the terminal settling velocity depends on various factors related to the particle and the fluid and is derived by balancing the forces acting on the particle (drag, buoyant and gravitational forces):

$$V_p \cdot g(\rho_s - \rho) = \frac{C_D \cdot u_{t,p}^2 \cdot \rho \cdot A_D}{2} \tag{1}$$

where: $u_{t,p}$ = terminal settling velocity of a particle with a diameter d_p [m/s]; V_p = effective volume of the particle [m^3]; g = gravitational constant [m/s^2]; ρ_s = particle density [kg/m^3]; ρ = fluid density [kg/m^3]; C_D = drag coefficient; A_D = projected area of the particle in the flow direction [m^2].

Then the terminal settling velocity is given by the following equation:

$$u_{t,p} = \sqrt{\frac{2 \cdot g(\rho_s - \rho) \cdot V}{C_D \cdot \rho \cdot A_D}} \tag{2}$$

in case of a solid and spherical particle:

$$u_{t,p} = \sqrt{\frac{4 \cdot g(\rho_s - \rho) \cdot d_p}{3 \cdot C_D \cdot \rho}} \tag{3}$$

where: d_p = particle diameter [m].

The terminal settling velocity is not depended on the horizontal and vertical movement of the fluid, although in real situation velocity gradients and other factor could affect the process [20], whereas the drag coefficient depends on the flow regime surrounding the particle. In the fluid dynamics the Reynolds number (Re) is used to identify the flow characteristics, thus by assuming the conditions previous described:

$$Re = \frac{d_p \cdot \rho \cdot u_{t,p}}{\mu} \tag{4}$$

where: μ = fluid viscosity [kg/m·s].

An extensive review of equations for C_D calculation is given by Brown and Lawler [21]. The drag coefficient was seen to decrease as the Reynolds number increases, sometimes a shape factor is also determined and incorporated in C_D to take into account the real particle contour. In case of spherical particles, the main equations for C_D and $u_{t,p}$ are given in Table 2 (extracted from Coulson et al. [20]). As indicated in the table, in the laminar regime (region a) the equation for the $u_{t,p}$ calculation is the Stokes' law and, in the turbulent regime (region c), it becomes the Newton's law.

Table 2. Drag and terminal settling velocity equations [20]

Regime	Re	C_D	$u_{t,p}$
Laminar (region a)	Re < 1	24/Re	$u_{t,p} = \dfrac{g \cdot d_p^2 \left(\rho_s - \rho\right)}{18 \cdot \mu}$ (Stokes)
Intermediate (region b)	1 < Re < 1000	24/Re + 0.44	$u_{t,p} = \sqrt{\dfrac{4 \cdot g\left(\rho_s - \rho\right) \cdot d_p}{3 \cdot C_D \cdot \rho}}$
Turbolent (region c)	1000 < Re < $2 \cdot 10^5$	0.44	$u_{t,p} = 1.75 \cdot \sqrt{\dfrac{g \cdot d_p \left(\rho_s - \rho\right)}{\rho}}$ (Newton)
Turbolent (region d)	Re > $2 \cdot 10^5$	0.10	$u_{t,p} = \sqrt{\dfrac{4 \cdot g\left(\rho_s - \rho\right) \cdot d_p}{3 \cdot C_D \cdot \rho}}$

If the settling velocity term is not known, to identify the range in which the motion of the particle lies, it must be eliminated from the Reynolds number. To this scope a criterion based on the K parameter was introduced [22], which is defined as:

$$K = d_p \left[\frac{g \cdot \rho \cdot \left(\rho_s - \rho\right)}{\mu^2} \right]^{1/3} \tag{5}$$

The Stokes' law is to apply if Re < 1, thus by substituting the proper $u_{t,p}$ in equation (4):

$$Re = \frac{d_p^3 \cdot g \cdot \rho \cdot (\rho_s - \rho)}{18 \cdot \mu^2} \tag{6}$$

$$Re = \frac{K^3}{18} \tag{7}$$

By solving equation (7), $K = 2.6$; this means that if K is less than 2.6 the Stokes' law applies.

Setting $Re = 1,000$ and $Re = 200,000$, after substituting the $u_{t,p}$ from the Newton's law:

$$Re = 1.75 \cdot K^{1.5} \tag{8}$$

By solving equation (8), K is respectively equal to 68.9 and 2,360.

The criterion to choose the proper equation for the calculation of the terminal settling velocity is summarized in Table 3. Figure 5 shows the trend of the terminal settling velocity as a function of the particle's diameter (assuming the ash density equal to 2000 kg/m^3). In the same graph the range of applicability, in term of K, is highlighted by means of the lines $K = 2.6$, $K = 68.9$ and $K = 2360$.

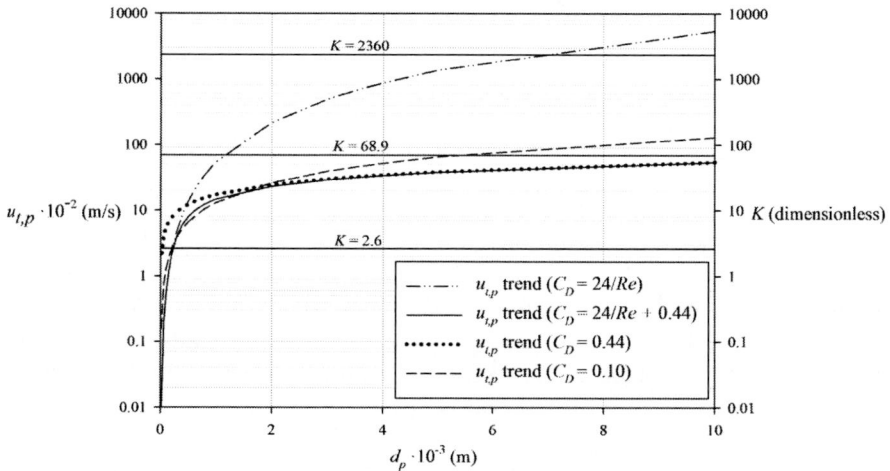

Figure 5. Terminal settling velocity as a function of the particle's diameter.

Table 3. K parameter and regime flow regions

Regime	Re	K
Laminar (region a)	Re < 1	$K < 2.6$
Intermediate (region b)	$1 < $ Re $ < 1000$	$2.6 < K < 68.9$
Turbolent (region c)	$1000 < $ Re $ < 2 \cdot 10^5$	$68.9 < K < 2{,}360$
Turbolent (region d)	Re $ > 2 \cdot 10^5$	$K > 2{,}360$

Within the approach mentioned above, drag coefficients and terminal falling velocities of Table 2 are based on the following assumptions:

- the settling is not affected by the presence of other particles in the fluid (*free settling*);
- the walls of the vessel do not exert an appreciable retarding effect;
- the effects of particle shape and orientation on drag are not accounted for.

Several studies were made to overcome such assumptions.

When the interference of other particles is appreciable, the process is known as *hindered settling*. Richardson and Zaki [23] stated that in concentrated suspension the drag force on a particle will be influenced by the concentration of particles and the terminal settling velocity will be a function of the voidage of the suspension as given below:

$$\frac{u_{t,p|C}}{u_{t,p}} = \left(1 - C\right)^n = e^n \tag{9}$$

where: $u_{t,p|C}$ = terminal settling velocity in hindered conditions [m/s]; C = suspension concentration expressed as volumetric fraction [dimensionless]; e = voidage [dimensionless]; n = empirical exponent dependent on Re.

Other subsequent studies were related to the quantification of the wall effects, these were taken into account by introducing some modifications in the n-value [24]; finally also the modelling of the sedimentation process of multisized particles was faced [25].

To take into account the effect of particle shape and its orientation on drag, two difficulties were remarked: the first is that infinite non-spherical shapes exist and the second is that each of these shapes is associated with an infinite number of orientations. Then, drag coefficients for generic non-spherical particles were defined in [26].

3.1. Condition for the Incomplete Particles Removal in Horizontal Flow Grit Chambers

Assume a rectangular settling vessel (Figure 6), where the sewage is fed (Q = sewage flow rate; L = vessel length; W = vessel width; H = vessel height). During the flow of the fluid, two zones can be distinguished, the suspension zone and the sludge deposit. The following assumptions were made:

- a homogenous stream, uniformly distributed over the tank cross-sectional area, is fed;
- the liquid in the feeding zone moves at constant velocity and as a plug flow;
- when particles enter the sludge zone (assumed at constant thickness), they exit the suspension.

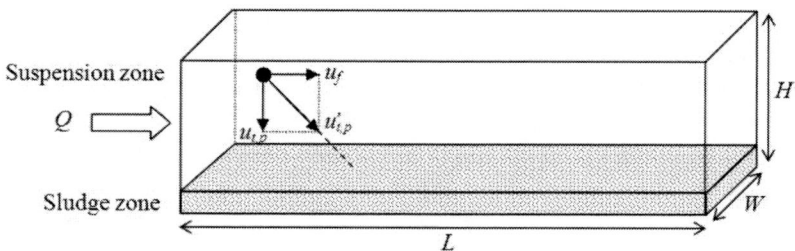

Figure 6. Rectangular settling vessel.

The final settling velocity of the particle ($u'_{p,t}$) is the vectorial sum of the terminal setting velocity ($u_{t,p}$) and the sewage velocity (u_f). In these conditions it is possible to define the *retention time* (t_o) and the *critical velocity of settling* for the particles (u_o). Particles with a settling velocity equal or lower than the critical velocity will settle out from the vessel at a time equal or greater than the retention time:

$$t_o = \frac{V}{Q} = \frac{L}{u_f} \tag{10}$$

$$u_o = \frac{H}{t_o} \tag{11}$$

It must be pointed that the design factors, mentioned above, must be adjusted to account for the effects of inlet and outlet turbulence, short circuiting, sludge storage and velocity gradient.

Then, assuming that particles are uniformly distributed over the entire depth (H) of the vessel at the inlet, the particles (X_r) with a terminal settling velocity less than u_o will be removed in the ratio:

$$X_r = \frac{u_{t,p}}{u_o} \tag{12}$$

In a typical suspension a large gradation of particle sizes occurs, thus to determine the removal efficiency it is necessary to consider the entire range of $u_{p,t}$ present in the system. This can be accomplished by using the results of a sieving analysis to construct a terminal settling velocity analysis curve. The total fraction of removed particles is given by:

$$x_{r,total} = \left(1 - X_o\right) + \int_o^{X_o} \frac{u_{t,p}}{u_o} dX \tag{13}$$

The term $(1 - X_o)$ is the fraction particles with a velocity greater than u_o, whereas $\int_o^{X_o} \frac{u_{t,p}}{u_o} dX$ is the fraction of particles removed with $u_{t,p}$ less than u_o.

3.2. Ash Characterisation

Some experimental tests are needed in order to apply the methodology described above and quantify the amount of volcanic ash which will be removed. The following experimental tests were performed on two samples of ash:

- Analysis of the particle size distribution;
- Density determination.

3.2.1. Analysis of the Particle Size Distribution

The analysis of the particle size distribution is made using sieves arranged in a column, in such a way that the top has the larger mesh and the others have a gradually smaller mesh going down to the bottom. The material is shaken or agitated above a mesh; particles of smaller size than the mesh openings can pass through them under the force of gravity. The sieving gives a series of particles classified into size ranges [27].

3.2.2. Density Determination

The density of a solid that does not dissolve in working liquid (water) can be determined by means of a pycnometer; the resulting density is the so-called *true density*.

The pycnometer (also called pyknometer or specific gravity bottle), is a flask with a close-fitting ground glass stopper with a fine hole through it, so that a given volume can be accurately obtained.

This enables the density of a fluid to be measured accurately, by reference to an appropriate working fluid such as water or mercury, by using an analytical balance [28].

The particle density of a powder, to which the usual method of weighing cannot be applied, can also be determined with a pycnometer. The material is added to the pycnometer, which is then weighed, giving the weight of the powder sample (by subtracting the weight of the pycnometer). The pycnometer is then filled with a liquid of known density, in which the powder is completely insoluble. Then weight of the displaced liquid can be determined and, finally, also the density of the powder (*solid density* or *true density*).

4. CASE-STUDY

The case study is the surrounding Mt. Etna in the South Italy (Figure 7). This volcano is one the highest volcanoes in Europe (~ 3300 m) and is characterised by both basaltic explosive behaviour and effusive activity.

Recently an increased trend to give explosive eruptions with ash emission has been observed (in particular during 2001, 2002–2003 and 2013) [29, 30]. In addition the expected damage caused by ash fallout is supposed to rise due to the increased urbanization and industrialization during recent years.

To apply the methodology described in Section 3 to the introduced case-study, two sample of volcanic ash produced by eruptive explosions of Mt. Etna were collected in the following points (see locations in Figure 7):

- sample ID = 1, collected close to the Cratere Silvestri (coordinates: lat. 37°41'55.73"N, long. 15°0'16.94"E; distance 5.5 km from the main crater)
- sample ID = 2, collected in urban area of Messina (coordinates: lat. 38°10'16.66"N, long: 15°31'25.56"E; distance 65 km from the main crater).

The samples of ash were characterised to determine the particle size distributions and the solid densities.

Then, the K parameter for the wastewater stream was calculated to make possible the choice of the approach for the computation of the terminal settling velocities.

The characteristics of the equipment, chosen for this study, are given in Table 4.

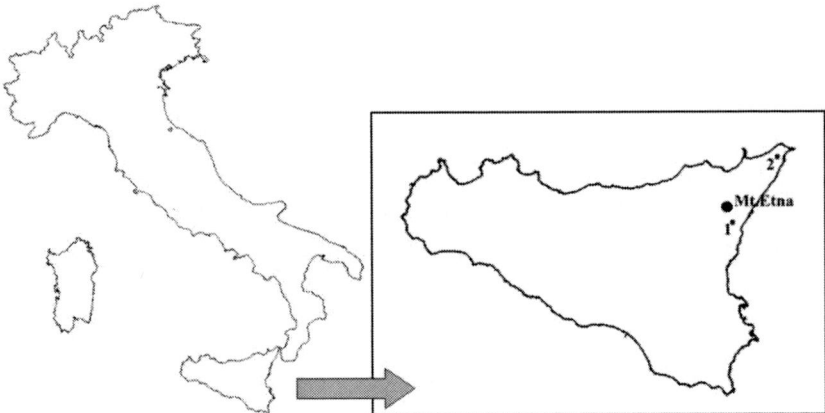

Figure 7. Location of Mt. Etna (Italy) and of the collection points.

Table 4. Horizontal grit removal chamber characteristics [31]

Parameter	Symbol	Unit	Value
Wastewater flow rate	Q	m^3/s	1.215
Wastewater velocity	u_f	m/s	0.3
Vessel length	L	m	18.00
Vessel width	W	m	3.00
Vessel height	H	m	1.35
Cross section	A	m^2	4.05
Retention time	t_o	s	60
Critical velocity of settling	u_o	m/s	0.0225

5. RESULTS AND DISCUSSION

The results of both the ash characterisation and the methodology application are given in the following Sections. Some comments are also provided.

5.1. Particle size Distribution

The particle size distributions of the ash samples are given in Table 5 (where w_i is the percentage remaining to the sieve with a given mesh), further details about the analysis of the particles size distribution with respect to these samples were given in [32, 10]. The prevailing diameters for the ash particles are 0.1 ÷ 2 mm for the sample 1 and 0.15 ÷ 2 mm for the sample 2.

Table 5. Particle size distribution for the samples of volcanic ash

Weight (%) $d_p \cdot 10^{-3}$ (m)	w_1 1.8	w_2 0.6	w_3 0.3	w_4 0.2	w_5 0.15	w_6 0.1	w_7 0.08	w_8 0.06
Sample 1	4.5	35.52	35.25	11.82	4.64	3.76	1.63	2.85
Sample 2	2.97	51.86	42.92	1.02	0.23	0.23	0.11	0.63

5.2. Density

The solid densities of both the samples were determined with the pycnometer, according to the procedure described in Section 3.2.2.

Results are given in Table 6.

Table 6. Ash densities

Sample ID	Solid density (kg/m^3)
1	2700
2	2000

5.3. Terminal Settling Velocity

To calculate the terminal settling velocities of the ash particles, the previous assumptions discussed in Section 3 were made and justified as follows: (1) given that the sewage can be considered a dilute suspension, the *free settling* condition occurs in the chamber; (2) the effect of the walls of the vessel is negligible as this facility is an industrial-scale equipment; (3) the particle shape is greatly variable to account for the infinite shapes, thus they need to be assumed spherical.

By applying the K parameter criterion, described above, the proper correlation for $u_{t,p}$ was chosen.

Then the terminal settling velocity was calculated for each representative diameter of the particle size classes, identified by sieving, and for both the samples (Table 7 and 8).

Figure 8 shows the trend of the settling velocity (left hand axis) as a function of particle diameter; in the right axis the K value is shown (in logarithmic scale) and allows delineating the regime regions listed in Table 3.

Table 7. Terminal settling velocity for the sample 1 of volcanic ash

$d_p \cdot 10^{-3}$ (m)	1.8	0.6	0.3	0.2	0.15	0.1	0.08	0.06
K	45.99	15.33	7.66	5.11	3.83	2.55	1.92	1.53
$u_{t,p} \cdot 10^{-2}$ (m/s)	28.68	13.45	6.21	3.31	1.98	0.93	0.52	0.33
$u'_{t,p} \cdot 10^{-2}$ (m/s)	41.5	32.9	30.6	30.2	30.1	30.0·	30.0·	30.0·

Table 8. Terminal settling velocity for the sample 2 of volcanic ash

$d_p \cdot 10^{-3}$ (m)	1.8	0.6	0.3	0.2	0.15	0.1	0.08	0.06
K	38.53	12.84	6.42	4.28	3.21	2.14	1.71	1.28
$u_{t,p} \cdot 10^{-2}$ (m/s)	21.67	9.56	4.02	2.03	1.19	0.55	0.31	0.20
$u'_{t,p} \cdot 10^{-2}$ (m/s)	37.1	31.5	33.0	30.1	30.0	30.0	30.0	30.0

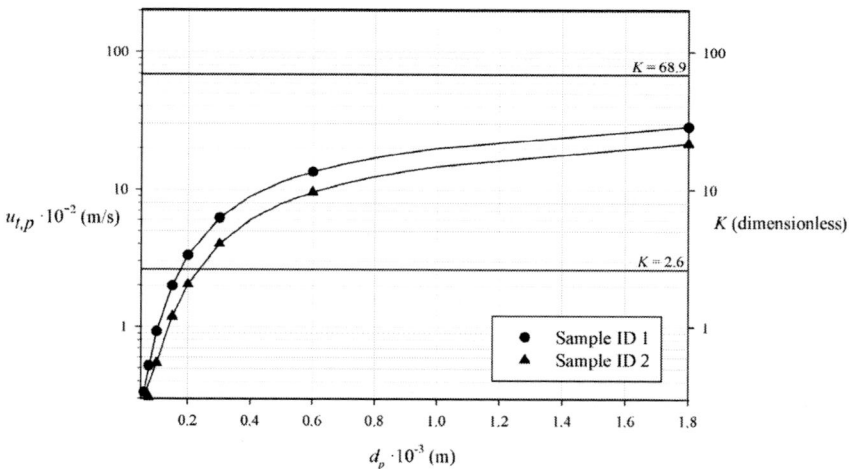

Figure 8. Terminal settling velocity of volcanic ash particles versus particles' diameter.

5.4. Fraction of Settled Particles

The weight fraction remaining to each sieve (X_w) of both the samples was plotted versus the terminal settling velocity as shown in Figure 9. Then, using the results of the analysis of the particle size distribution, the weight fraction having a velocity of settling less than $u_{t,p}$, which is the fraction passing each sieve (X), was calculated.

Figures 10 and 11 give the curves representing the weight fraction versus the terminal settling velocity, respectively, for the sample 1 and 2; the value of u_o is indicated in both the figures.

After deriving the equation for the curves of Figure 10 and 11 by means of a regression procedure, the total fraction of removed particles was calculated for both the sample by using equation (13).

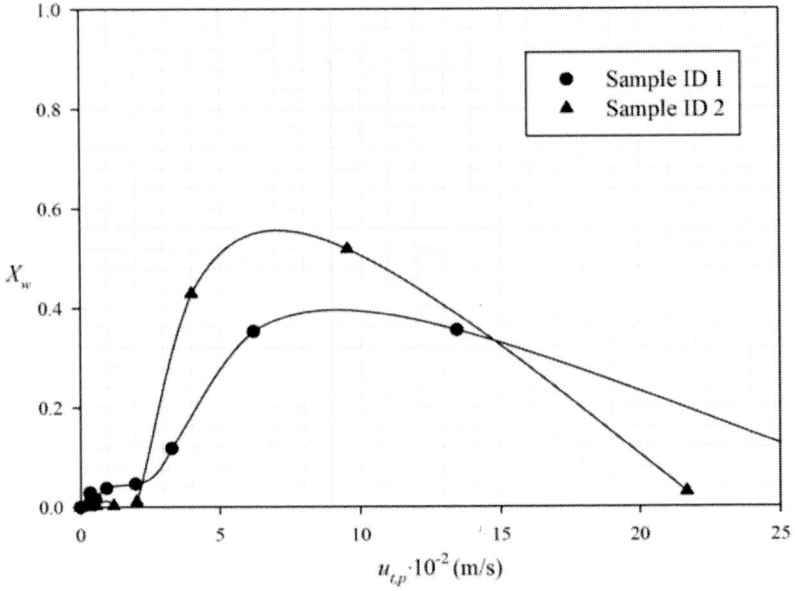

Figure 9. Weight fraction remaining to each sieve versus terminal settling velocity of particles.

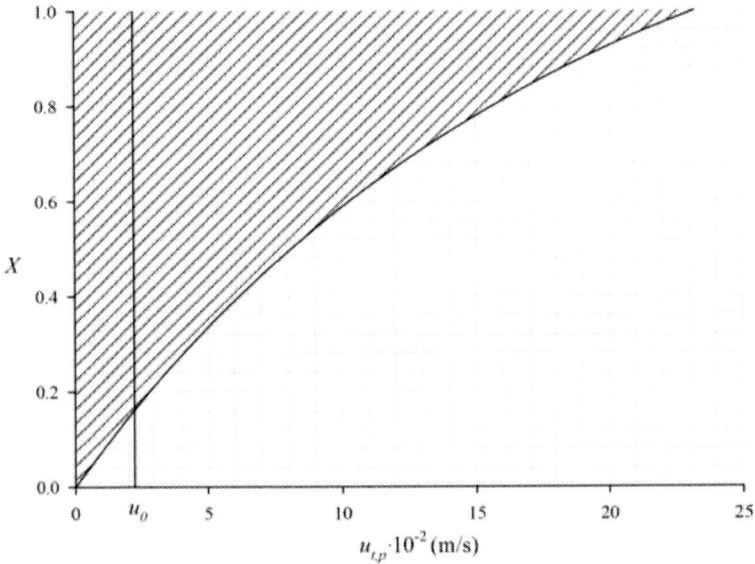

Figure 10. Fraction of ash with settling velocity less than $u_{t,p}$ versus settling velocity (sample 1).

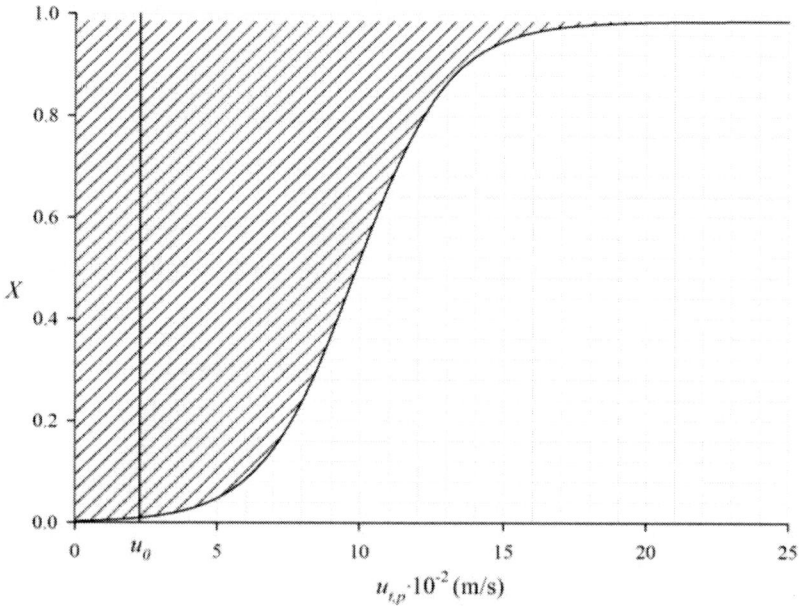

Figure 11. Fraction of ash with settling velocity less than $u_{t,p}$ versus settling velocity (sample 2).

The unsettled fraction is given by the area under the curves on the left hand of the line passing through u_o. The results of the total fraction of removed particles are given in Table 9.

Table 9. Fraction of removed particles

Sample ID	X_o	$1 - X_o$	$\int_{o}^{X_o} \dfrac{u_{t,p}}{u_o} dX$	$X_{r,total}$
1	0.175	0.825	0.089	0.914
2	0.011	0.989	negligible	~ 0.999

DISCUSSION

The samples of volcanic ash produced by Mt Etna, used in this study, are the same ones considered by Milazzo et al. (2014) [10]; thus the separation effectiveness of the grit removal equipment could be compared with that related to the use of a fine screen.

Firstly, some considerations concerning the particles of ash contained in the samples can be made: (*i*) sample 1, collected at ~ 5 km from the crater, contains a predominant percentage (~ 70%) of particles whose diameter is between 0.6 and 0.2 mm; (*ii*) sample 2, collected at ~ 65 km from the crater, contains a predominant percentage of particle of ~ 0.6 mm (~ 52%). The determination of the solid density (*true density*) justifies the results of the particle size distributions: in fact, by referring to particles of the same diameter (i.e. 0.6 mm), the higher density of the sample 1 causes the fallout of particles with the selected size much more at the crater proximity compared to sample 2. Obviously this is not the only factor affecting the fallout distances and the weighted percentage per each particle size, this issue is not the object of this work.

In calculating the terminal settling velocity, it was observed that the K parameter is always smaller than 68.9. The particle sizes do not allow the achievement of the K values required to apply the Newton's law ($68.9 < K < 2,360$) and the correlation associated with the region d ($K > 2,360$). As shown in Section 3, the density values influence the terminal settling velocity. By comparing particles with the same diameter of both the samples, particles of the sample 1 have a greater sedimentation velocity than those of the sample 2. Figure 9 gives an immediate view of the weighted fraction of particles for each possible value of terminal settling velocity. The results of the particle size distribution analysis allowed drawing the curves of the fraction of particles having a terminal settling velocity lower than a stated velocity $u_{t,p}$ for both the samples. Then, by computing the critical velocity u_o for particles in the grit chamber, the fraction of removed particles has been quantified. Results showed that ~ 94 % of particles of the sample 1 are removed, whereas particles of the sample 2 can be considered totally removed.

Additional considerations can be made on the choice of the design parameters for the equipment or the formulation of potential alternatives to horizontal grit chambers. Given that wastewater plants are designed to remove about the 95% of particles with a diameter of 0.21 mm, it can be observed that: in the case of sample 1 the $u_{t,p}$ is $3.3 \cdot 10^{-2}$ m/s for particles with $d_p = 0.2$ mm, this means that there is a no-negligible fraction of particles having $u_{t,p} < u_o$; in the case of sample 2, the $u_{t,p}$ of particles of the same diameter is $2.03 \cdot 10^{-2}$ m/s that is greater than u_o. The critical settling velocity can be increased only by increasing the depth of the chamber, since a reduction of t_o bring out from the range of the design values for u_f and L. Thus, in this frame the only possible alternative is the use of vortex-type grit chambers, as they remove particles up to 0.11 mm with a retention time of 30 s, but unfortunately their use increase costs; aerated grit chambers are not recommended because they eliminate particle up to 0.20 mm and hence, on the basis of the results of this study, they do not introduce any significant improvement in the removal process compared to a horizontal grit chamber.

CONCLUSION

In this chapter the conditions leading to failures or malfunctions of grit removal facilities with the respect to the phenomenon of volcanic ash emission, were determined. The work was mainly addressed to the investigation of horizontal grit chambers for a specific case-study, which is the surrounding of Mt. Etna.

Some assumptions were necessary for the application of the methodology discussed in Section 3. All these assumptions were proper discussed and justified and, even they were necessary to simplify the application of the approach, the results gave a valid support in addressing alternative solutions for the ash removal, such as the use of a vortex-type grit chamber even if it is much more expensive compared to a horizontal type.

REFERENCES

[1]   Fabiano, B.; Currò, F. *Process Saf. Environ.* 2012, 90, 357-367.
[2]   Antonioni, G.; Bonvicini, S.; Spadoni, G.; Cozzani, V. *Reliab. Eng. Syst. Safe* 2009, 94, 1442-1450.
[3]   Ancione, G.; Salzano, E.; Maschio, G.; Milazzo, M. F. *J. Risk Res.* 2014, in press, doi: 10.1080/13669877.2014.961515.
[4]   Kadri, F.; Birregah, B; Châtelet, E. *J. Homeland Security Emerg. Manag,* 11(2), 217-241.
[5]   Girgin, S.; Krausmann, E. *J Loss Prevent Proc* 2013, 26, 949-960.
[6]   Cruz, A. M.; Steinberg, L. J.; Arellano, A. L. V.; Nordvik, J. P.; Pisano, F. State of the art in Natech risk management. Report no. 21292, *European Commission Directorate General Joint Research Centre,* EUR, 2004.

[7]   International Civile Aviation Organization 2012. Flight Safety and Volcanic Ash. Risk management of flight operations with known or forecast volcanic ash contamination Document no. 9974 ANB/487 (1st edition) [available online: http://www.icao.int/publications; accessed 30.09.2014]

[8]   Zais, D. May 18, 1980: The Day the Sky Fell Managing the Mt. St. Helens Volcanic Ashfall on Yakima, Washington, U.S.A. [available online: www.volcanoes.usgs.gov; accessed 06.08.2014].

[9]   Ancione, G.; Salzano, E.; Maschio, G.; Milazzo, M. F. Chem Eng Trans 2014, 36, 433-438, doi: 10.3303/CET1436073.

[10] Milazzo, M. F.; Ancione, G.; Primerano, P.; Salzano, E.; Maschio, G. The impact of volcanic ash fallout on industrial facilities: Natural-Technological hazards in wastewater treatments (Screening Processes). Volcanic Eruptions: Triggers, Role of Climate Change and Environmental Effects; Nova Publication, in press.

[11] Baxter, P. J.; Bernstein, R. S.; Falk, H.; French, J.; Ing, R. *Disasters,* 1982, 6, 268-276.

[12] Spence, R. J. S.; Baxter, P. J.; Zuccaro, G. *J. Volcanol. Geoth. Res.* 2004, 133, 321-343.

[13] Rasà R.; Tripodo A.; Casella S.; Szilagyi M. L. Contributi alla valutazione della pericolosità rischio vulcanico nell'area etnea ed alla mitigazione dei danni attesi. Carta Tematica di Rischio Vulcanico della Regione Sicilia Ed. C.R.P.R. - *Regione Sicilia Palermo*, 2007, (in Italian).

[14] Milazzo, M. F.; Ancione, G.; Basco, A.; Lister, D.; Salzano, E.; Maschio, G. *Nat Hazards* 2013, 66, 939-953, doi: 10.1007/s11069-012-0518-5.

[15] Milazzo, M. F.; Ancione, G.; Lister, D. G.; Basco, A.; Salzano, E.; Maschio, G. *Chem. Eng. Trans.* 2012, 26, 123-128, doi: 10.3303/CET1226021.

[16] Milazzo, M. F.; Ancione, G.; Salzano, E.; Maschio, G. *Reliab Eng Syst Safe* 2013, 120, 106-110, doi: 10.1016/j.ress.2013.05.008

[17] Metcalf & Eddy, Inc. Wastewater engineering Treatment, Disposal, and Reuse, (3rd ed.); Irwin/McGraw-Hill: Boston, US 1991; pp 1-13.

[18] USEPA 2003. Wastewater Technology Fact Sheet. Screening and Grit Removal. Document no. EPA832-F-03-011 [available online: http://www.epa.gov/ owm/mtb/ mtbfact.htm; accessed 06.08.2014].

[19] Water Environment Federation, 2007. Operation of Municipal Wastewater Treatment Plant. WEF Manual of Practice no. 11 [available online: http://www.wefnet.org; accessed 30.09.2014].

[20] Coulson, J. M.; Richardson, J. F.; Backhurst, J. R.; Harker, J. H. Chemical engineering: Particle technology and separation processes, (5th ed.); Butterworth-Heinemann: Oxford, UK 2002; Vol. 2, pp 191-203.

[21] Brown, P. P.; Lawler, D. F. J Environ Eng ASCE 2003, 129, 222-231.

[22] Zhang, J. Motion of Particles through Fluids. Study Notes, Chapter 2. [available online: http://lorien.ncl.ac.uk/ming/particle/cpe124p2.html]

[23] Richardson, J. F.; Zaki, W. N. *Trans Inst. Chem. Eng.* 1954, 32, 35-53.

[24] Cheng, N. S. *J. Hydraul. Eng. ASCE,* 1997, 123(8), 728-731.

[25] Kothari, A. C. Sedimentation of Multisized Particles. *Thesis in Chemical Engineering.* 1981 Texas Tech University.

[26] Clift, R.; Grace, J. R.; Weber, M. E. Bubbles, *Drops and Particles.* Academic Press Inc: London, UK 1978; pp 142-168.

[27] Allen, T. Powder sampling and particle size determination (1st ed.). Elsevier: Amsterdam, The Netherland 2003; pp 208-250.

[28] ASTM D854. Standard Test Methods for Specific Gravity of Soil Solids by Water Pycnometer.

[29] Global Volcanism Program web-site [http://www.volcano.si.edu/; accessed 07.08.2014].

[30] Branca, S.; Del Carlo, P. Eruptions of Mt. Etna during the past 3,200 years: a revised compilation integrating the historical and stratigraphic records. "Mt. Etna: Volcano Laboratory" *Geophys. Monogr.* 2004, 143, 1-27.

[31] Liu, D. H. F.; Liptak B. G. *Environmental Engineers' Handbook* (2nd ed.). CRC Press: Priceton, New Jersey, 1997; pp 658-664.

[32] Milazzo, M. F.; Primerano, P.; Ancione, G.; Salzano, E.; Maschio, G. *Chem. Eng.Trans.* 2014, 36, 487-492; doi: 10.3303/CET1436082.

In: Volcanic Ash
Editor: Danielle Graver

ISBN: 978-1-63463-597-4
© 2015 Nova Science Publishers, Inc.

Chapter 3

A REVIEW OF TEPHRA TRANSPORT AND DISPERSAL MODELING STRATEGIES AND APPLICATIONS TO FAR-RANGE HAZARD ASSESSMENT AT SOME OF THE MOST ACTIVE VOLCANOES IN THE WORLD

Bosanna Bonasia[1] and Chiara Scaini[2]*

[1]Centro de Geociencias, Universidad Nacional Autónoma de México, Campus Juriquilla, Querétaro, México.
[2]Barcelona Supercomputing Center-Centro Nacional de Supercomputación, Barcelona, Spain

ABSTRACT

Ash produced as consequence of explosive volcanic eruptions can cause multiple hazards both close to the volcano and at great distances. Ash fallout can produce a wide range of impacts on exposed assets (edifices, roads, lifelines), while airborne ash jeopardizes air travel safety. The attention on the impacts of ash dispersal on air traffic increased after the severe consequences of the 2010 Eyjafjallajökull eruption (Iceland), which caused an unprecedented closure of the European North Atlantic airspace. Thus, the need to predict the dispersal of volcanic ash became more important for both civil-defense authorities and airline stakeholders

---

* Corresponding author: rbonasia@geociencias.unam.mx.

(companies, service providers). For this reason, tephra transport and dispersal models, which allow predicting tephra sedimentation and dispersal caused by a modelled eruptive scenario, became crucial in volcanic hazard and risk assessment. These models have been used for multiple purposes: short-term operational forecast of volcanic ash dispersion; reconstruction of past eruptions in order to constrain key eruption parameters; and generation of probabilistic hazard maps for hazard assessment of tephra dispersion and fallout. This chapter reviews the applications to long-term hazard assessment of volcanic ash dispersal and deposition for multiple eruptive scenarios at some of the most active volcanoes in the world. We present the improvements achieved during the last two decades focusing on recent developments of tephra dispersal hazard assessment, and we discuss some relevant aspects that must be improved in the future in order to optimize long-term hazard assessment strategies.

1. INTRODUCTION

Eruptive columns produced by explosive volcanic eruptions eject a high quantity of pyroclastic particles into the atmosphere. Tephra[1] fallout can affect large areas, but is mainly constrained to proximal and medial distances from the source (tens to few hundreds of km). The finer fraction of tephra can be transported thousands of km away from the volcano, dispersed under the combined effect of wind advection and atmospheric turbulence.

Substantial ash fallout can cause collapse of edifice roofs (Spence et al., 2005), while moderate ash fallout can disrupt electricity networks (Wardman et al., 2012), lifelines (Wilson et al., 2011) and water resources (Stewart et al., 2006; 2009). Ash fallout and remobilization can also affect public health, causing, among others, respiratory problems (Horwell and Baxter, 2006). In addition, the accumulations of a few millimeters of ash can disrupt airport operations (Casadevall, 1993; Guffanti et al., 2009). Finally, ash deposition can also trigger secondary hazards, such as, for example, lahars (Favalli et al., 2006), that should be taken into account in a comprehensive hazard assessment (Paton et al., 2001). Thus, ash fallout can be the cause of disruption of a wide range of human activities, leading in some cases to strong socio-economic impacts (Zuccaro et al., 2008; Zuccaro et al.,

---

[1] *Tephra* is the general term for the fragmented material ejected during an explosive volcanic eruption, regardless of its size and composition. The term *ash* refers only to the finest fraction of tephra, i.e. Particles smaller than 2 mm in diameter.

2013; Jenkins et al., 2014; Scaini et al., 2014). Besides the impact of tephra fallout, ash dispersal in the atmosphere poses a high threat to air navigation (Casadevall, 1993; Miller and Casadevall, 2000). Ash clouds can have a hemispheric impact, as demonstrated during the eruption of the Icelandic volcano Eyafjallajökull, which occurred in April-May 2010 and caused an unprecedented closure of the European and the North Atlantic airspace, with global economic losses of million of dollars. In addition, the recent eruption of Cordón del Caulle in Chile (2011) caused strong impacts on aviation and led to blockage of air traffic in the affected area. Thus, in recent years, impacts of volcanic ash on civil aviation have become a global issue to deal with, in particular due to the strong increase of passenger numbers and the growing number of civil airports.

It is thus necessary to improve short-term prediction of ash dispersal in the atmosphere, in order to provide useful information in the unrest and emergency phase. Quantitative ash dispersal forecasts, introduced during the 2010 Eyafjallajökull eruption (Bolic and Sivcev, 2011), are based on numerical simulations performed with Tephra Transport and Dispersion Models (TTDM). Modeling results can support response strategies during the emergency and, if based on *a priori* eruptive scenarios, such as, for example, the operational method implemented for forecasting Mt. Etna volcanic plumes (Scollo et al. 2009), can also provide a preliminary assessment with hours/days advance notice.

Besides the importance of short-term emergency management, comprehensive risk management also requires long-term planning, aimed at assessing expected outcomes on a statistical basis. Medium to long-term analyses have become of fundamental importance for aviation stakeholders after the introduction of the Safety Risk Assessment (SRA), a framework applied by aviation stakeholders for volcanic hazards (ICAO, 2012). Airlines are nowadays able to perform operations in ash-contaminated airspaces, after having defined a SRA and having received the approval from the authority. Such procedure should rely on many aspects and, between others, the assessment of long-term expected conditions, that is, ash dispersal patterns more likely to happen.

To this end, probabilistic hazard maps allow assessing the expected impacts based on most probable scenarios and representative conditions. Probabilistic hazard assessment was first introduced for tephra fallout (Bonadonna et al., 2005; 2006) and is based on the definition of eruptive scenarios based on a wide range of Eruptive Source Parameters (ESPs) characterized by a probability distribution and defined based on the historical eruptive record. Modeling results are then merged to produce probability maps that account for the probability of overcoming certain hazardous conditions (Folch and Sulpizio, 2010). This methodology allows production of probabilistic hazard maps for ash dispersal, which are extremely useful for decision-makers and stakeholders involved in civil aviation management during explosive volcanic eruptions (Bonadonna et al., 2013).

Before 2010 there were only few examples of hazard assessment for tephra dispersal (i.e. Papp et al., 2005; Folch and Sulpizio, 2010), due to the high computational cost of performing many simulations (Folch, 2012). In recent years, the improvement of TTDM strategies and the introduction of novel techniques to tackle the computational costs, allowed the production of several hazard assessments for tephra dispersal at many active volcanic areas (Scaini et al., 2012; Sulpizio et al., 2012; Bonasia et al., 2014; Biass et al., 2014).

In this chapter we review some long-term applications of TTDM models for the assessment of hazards posed by tephra fallout and dispersal for some active volcanoes in the world (i.e. Tarawera, New Zeland; Vesuvius, Italy; Concepción, Nicaragua; and Popocatépertl, Mexico). We discuss the importance of a probabilistic approach in the volcanic ash hazard assessment and describe the peculiarities of the approaches used in the different cases. In fact, besides adapting existing tephra fallout hazard assessment techniques, some of these works introduce novel methodologies focused on different spatial and temporal scales. We discuss limitations of the presented strategies, and provide insights for future developments in this field.

2. STATE OF THE ART OF TEPHRA TRANSPORT AND DISPERSAL MODELS

In the last decades, the scientific community developed several models for simulating tephra ejection, transport and sedimentation during explosive volcanic eruptions (Folch, 2012). Examples of such models are HAZMAP (Macedonio et al., 2005; Pfeiffer et al., 2005), ASHFALL (Hurst and Turner, 1999; Hurst and Smith, 2004) and TEPHRA (Bonadonna et al., 2005), which can all be grouped in the category of Gaussian Tephra Transport and Dispersion Models. These models admit an analytical solution for the Advection-Diffusion-Sedimentation (ADS) equation, generally expressed as:

$$\frac{\partial C}{\partial t} = -\nabla \cdot uC + \nabla \cdot K\nabla C - \nabla u_s C + S_0 \tag{1}$$

where t denotes time, C is the particle mass concentration, $u=(u_x, u_y, u_z)$ is the wind velocity, K is the turbulent diffusivity tensor, u_s is the particle settling velocity, and $S_0(x,y,z,t)$ is the source term that accounts for the production and distribution of particles. The first three terms on the right hand side of Eq. (1) describe the advection of particles by wind, turbulent diffusion of particles and particle sedimentation, respectively. Eq. (1) is solved using a semi-analytical solution as described in Macedonio et al. (2005) and Pfeiffer et al. (2005). Usually a purely empirical description that reproduces the geometrical shape of the eruption column is adopted. Then, the source term in Eq. (1) is described as proposed by Suzuki (1983):

$$S x,y,z,t = S_0 \left(1-\frac{z}{H}\right)\exp\left[A\left(\frac{z}{H}-1\right)\right]^\lambda \times \delta\ t-t_0\ \delta\ x-x_0\ \delta\ y-y_0 \tag{2}$$

where S_0 is the normalization constant, x_0, y_0 are the coordinates of the vent, δ is the Dirac's distribution, which considers a filiform and instantaneous release, H is the column height, and A and λ are two empirical parameters that determine the position of the maximum concentration (located at $H(A-1)/A$) and how closely the mass is concentrated around the maximum (λ).

Gaussian TTDMs have low computational cost and are suitable to solve inverse problems, in order to constrain eruptive parameters (e.g. column height, total mass) from past eruptions and elaborate ground-load probabilistic hazard maps. However, because of the simplifying assumptions that characterize these models, they have some limitations under which they should not be applied (Folch, 2012).

For example they cannot simulate transport and deposition of fine ash because they cannot consider wind fields at large scales, nor the change of wind distribution in time. Moreover, Gaussian models cannot simulate dispersal from low and weak eruptive columns because they do not take topographic effects and wind interaction into account.

Fully numerical TTDMs are in turn able to simulate any kind of eruption and particle size, and solve the complete ADS equation, but their adoption implies a higher computational cost. Eulerian TTDMs allow calculating the time-dependent airborne ash concentration on a 3D computational grid.

An example of Eulerian TTDM is FALL3D (Costa et al., 2006; Folch et al., 2009), which solves the ADS equation (Eq. 1) on a structured terrain-following grid using a second-order Finite Differences scheme, with an atmospheric turbulent diffusion given by the gradient transport theory, semi-empirical particle terminal velocity parameterization and a time-dependent three-dimensional wind field. Application of TTDMs, such as FALL3D, includes the production of an operational forecast (Folch et al., 2008a; 2008b), modeling of past events (Costa et al., 2012), and hazard assessment (e.g. Scaini et al., 2012).

Finally, Lagrangian models (e.g. PUFF; Searcy, 1998) use particle tracers for the forecasting and computation of trajectories of ash particles in the atmosphere.

TTDMs rely on two types of inputs: meteorological and volcanological (e.g. Mastin et al., 2009). The main meteorological variables that influence tephra dispersal are wind speed and direction, relative humidity, temperature, pressure and atmospheric turbulence. The choice of the meteorological driver for TTDM simulations depends on the scale at which the simulation is performed, given that the two scales should be similar (Folch et al. 2012). At a global/regional scale, one may use the global meteorological datasets (e.g. NOAA NCEP/NCAR Reanalysis dataset, Kalnay et al 1996), available online, while if the simulation is performed at a local scale, it is necessary to run a Numerical Weather Prediction Model (NWPM) to produce a meteorological dataset at higher spatial resolution. In this case, TTDMs are coupled (usually off-line) with the NWPM, increasing the computational cost of the hazard assessment.

Finally, volcanological inputs are basically the input parameters related to the characteristics of the eruptions (e.g. column height, eruption duration, grain size distribution) and their definition is a critical issue for modelers due to the complexity of the phenomena at stake (Mastin et al., 2009). Given that the final aim of tephra dispersal hazard assessment is to assess the expected outcomes at active volcanic areas, their results should have a long-term validity.

Hazard assessment strategies must therefore account for these aspects in the selection of TTDM inputs representative of the long-term conditions.

3. TEPHRA FALLOUT AND DISPERSAL HAZARD ASSESSMENT OF DIFFERENT ERUPTIVE SCENARIOS AND MULTIPHASE ERUPTIONS

3.1. A Multiphase Eruption at Tarawera Volcano, New Zealand

Tarawera is one of the most destructive of New Zealand's volcanoes with its recent AD 1886 Plinian eruption that buried seven villages and caused more than 150 fatalities.

Bonadonna et al. (2005) described the probabilistic hazard assessment of tephra dispersion and accumulation for this volcano based on its recent Plinian eruption (AD 1315 Kaharoa eruption, Nairn et al., 2004). For the probabilistic modeling of this scenario, the authors developed and applied an analytical advection-diffusion TTDM (TEPHRA) that can run in parallel on multiple processors, and allows for grain-size dependent diffusion, a stratified atmosphere and particle diffusion time within the rising plume (Bonadonna et al., 2005). Different eruptive vents characterize the Tarawera volcano, and its eruptive history involves multi-phase events. In order to account for the complexity of eruptive scenarios, Bonadonna et al. (2005) introduced the definition of eruptive scenarios based on ranges of eruptive source parameters (ESPs). For the hazard assessment, they define three different eruptive scenarios: upper limit scenario (ULS), eruption range scenario (ERS) and multiple eruption scenario (MES). ULS relies on the highest column and the longest duration, while ERS and MES are defined based on different eruptions from the eruptive history. In the case of ERS and MES, the TTDM model is then initialized with a set of ESPs stochastically sampled within the scenarios considered. This novel modeling strategy allows forecasting a range of possible scenarios, and subsequent tephra fallout, and produces results that account for the wide variability (and uncertainty) associated with ESPs.

Figure 1 shows the probability maps for the ULS (a) and for the ERS (b), where a 26 km plume height and a plume ranging between 14 and 26 km, respectively, are considered. In both cases maps show that the main populated towns northeast of Tarawera would have a moderate to high probability of being affected by an ash loading of more than 10 kg/m^2 responsible for damage to vegetation.

In case of the occurrence of MES (Figure 2) the same towns Northeast of the volcano would have between a 90 and 100% probability of exceeding the defined ash-loading threshold, with the consequence that some of them could also experience collapse of the weakest buildings.

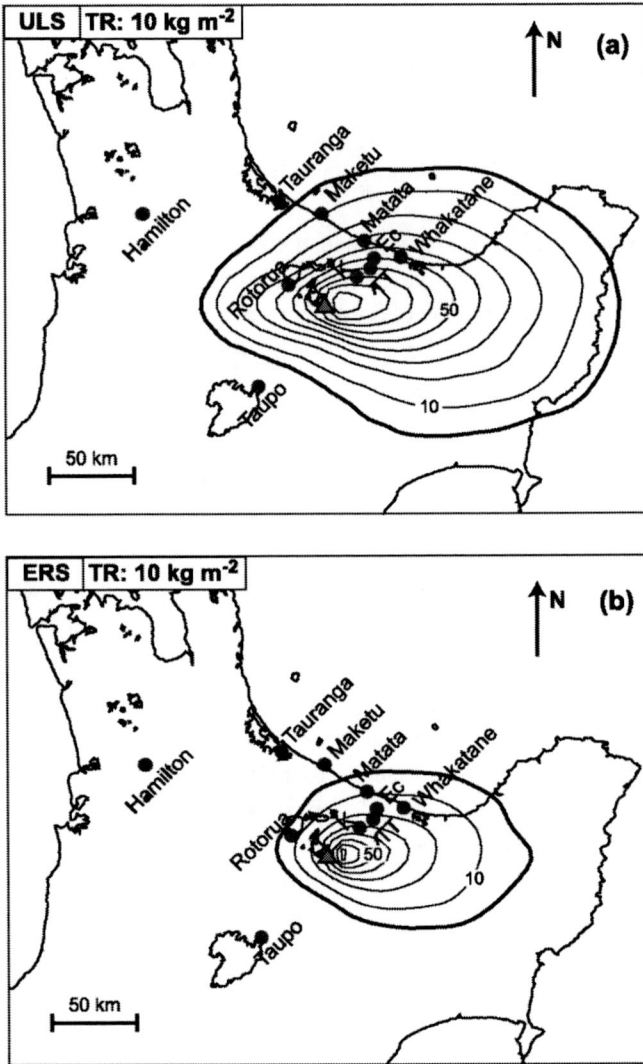

Figure 1. Probability maps for (a) upper limit scenario and (b) eruption range scenario. Contours are spaced for every 10% probability of reaching the threshold of damage to vegetation (i.e., 10 kg/m^2). The 5% contour is also shown (thick solid line). Key cities and towns are indicated with circles, and the Tarawera Volcanic Complex is indicated with a triangle. (After Bonadonna et al., 2005).

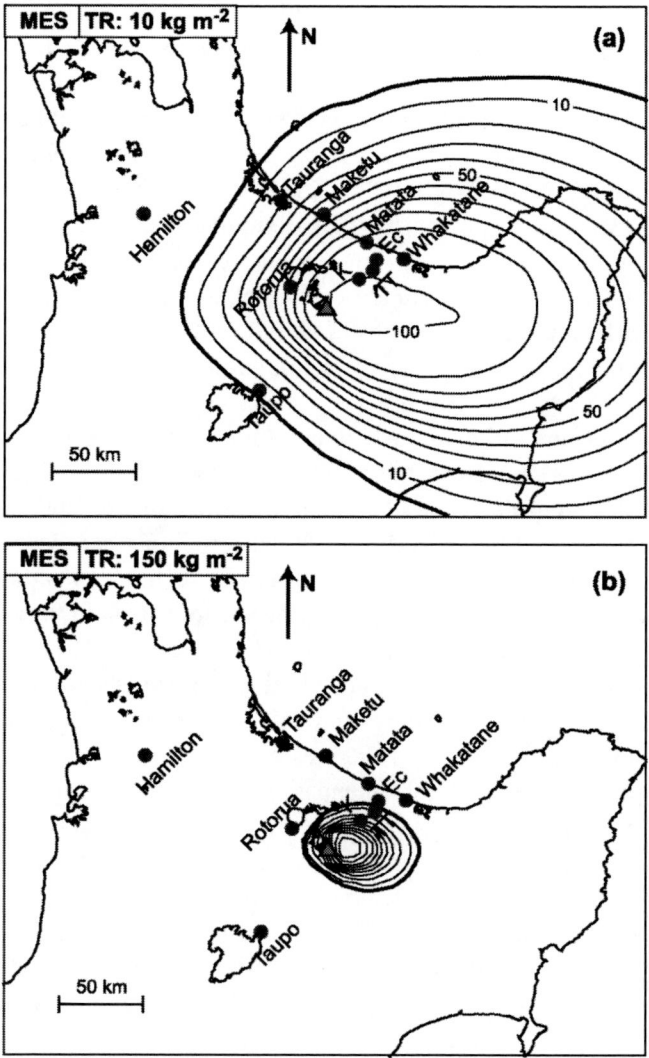

Figure 2. Multiple eruption scenario maps computed for a deposit threshold of (a) 10 kg/m² (damage to vegetation) and (b) 150 kg/m² (minimum loading for roof collapse). (After Bonadonna et al., 2005).

3.2. The Case Study of the Vesuvius Volcano (Italy)

During its eruptive history, Vesuvius experienced a long rest period often followed by very intense eruptions. The last eruption occurred in March 1944 and since that time the volcano is quiescent. Vesuvius is considered a potentially dangerous volcano due to the presence of strongly urbanized areas in its surroundings.

In the last decades, hazards associated with the possible renewal of explosive activity at Vesuvius have been the subject of several studies (e.g. Macedonio et al., 1988; Barberi et al., 1999; Cioni et al., 2003; Macedonio et al., 2008; Folch and Sulpizio, 2010). Its eruptive history is characterized by very different eruptive styles. Strombolian eruptions (e.g. the 1944 AD eruption), relatively low-magnitude events, characterized by the injection of incandescent lapilli and lava bombs to altitudes of tens to hundreds of meters, caused serious problems to local communities and in particular to the urbanized area of Naples. But the most severe ash fallout hazard is associated to Sub-Plinian and Plinian eruptions. Among the Plinian eruptions of Vesuvius, the most famous and extensively studied one is the 79 AD eruption, well known for the destruction of Roman towns of Pompeii and Herculaneum. Tephra fallout from Plinian and Sub-Plinian eruptions can cause substantial damages to the exposed population and assets: tephra deposition can cause roof collapse (Spence et al., 2005) and disruption of main services such as transportation and electricity networks (Wilson et al., 2011). Besides the impact of ash fallout, even low concentrations of ash at critical flight levels, can affect the safety of landing and take-off operations since it reduces visibility and can affect aircraft engines and components on both short and long-term periods. Hazard assessment is thus necessary in this area for both tephra fallout (due to its consequences at local/regional scale) and dispersal, which can affect the air traffic activities at regional to global scales.

Macedonio et al. (2008) assessed the relative hazard of ash fallout for three different possible scenarios at Vesuvius: Plinian, Sub-Plinian and violent Strombolian eruptions. Three reference eruptive scenarios were defined based on the 79 AD, the 1631 AD and the 1944 AD Vesuvius eruptions, respectively. Input eruptive parameters for the Plinian scenario are based on the best-fit parameters obtained by Pfeiffer et al. (2005) and shown in Table 1. For the Sub-Plinian scenario, input parameters were found by best-fitting field deposits data from Cioni et al. (2003) (Table 1). Finally, the violent Strombolian scenario, reconstructed on the base of the 1944 AD eruption, was modelled using input eruptive parameters derived from direct observation of column heights during the fire-fountaining episodes and through a semi-quantitative reconstruction of the event by using the FALL3D model (Table 1).

Fallout deposits for the first two scenarios were modelled using HAZMAP (Macedonio et al., 2005), which is based on the semi-analytical solution of the 2D ADS equation (Eq. 1). The meteorological inputs for hazard assessment (i.e. vertical wind profiles) were extracted from the NCEP/NCAR daily-averaged global meteorological dataset (Kalnay et al., 1996) at the point of the mesh grid nearest to Naples. The violent Strombolian scenario, characterized by low eruptive columns, was modelled by means of the fully numerical model FALL3D (Costa et al., 2006; Folch et al., 2009). This model adopts realistic near surface wind fields (i.e. meteorological data in gridded format) that account for terrain effects and uses a realistic evaluation of the turbulent atmospheric diffusion.

Results of these hazard assessments are probability maps for ash loading threshold values of 300 and 400 kg/m^2, which are considered critical for roof collapse of low and medium quality buildings respectively at the Neapolitan area (Zuccaro et al., 2008). Ground-load probability maps for the Plinian and Sub-Plinian scenarios (Figs 3 and 4) show that these two scenarios would affect a wide area where more than one million of people live nowadays. Figure 5 shows the ground-load probability maps that the authors obtained for the violent Strombolian event. Results show that the area subject to roof collapse in this case is encompassed by the hazard area identified by the Sub-Plinian scenario.

Table 1. Parameters used to obtain ground load probability maps for the three different scenarios (modified after Macedonio et al., 2008)

Model	Plinian	Sub-Plinian	Violent Strombolian
	HAZMAP	**HAZMAP**	**FALL3D**
Diffusion coefficient (m^2/s)	5000	5000	Computed by the model
Average column height (km)	27	18	4
Column shape parameters	A=4, λ=1.5	A=3, λ=1	BPT[a]
Average mass flow rate (kg/s)	8×10^7	3×10^7	5×10^5
Duration (h)	7	4.5	110
Total mass (kg)	2×10^{12}	5×10^{11}	2×10^{11}
Meteorological set	NCEP/NCAR	NCEP/NCAR	NCEP/NCAR

[a] Estimated by the model using the Buoyant Plume Theory (BPT, Bursik, 2001).

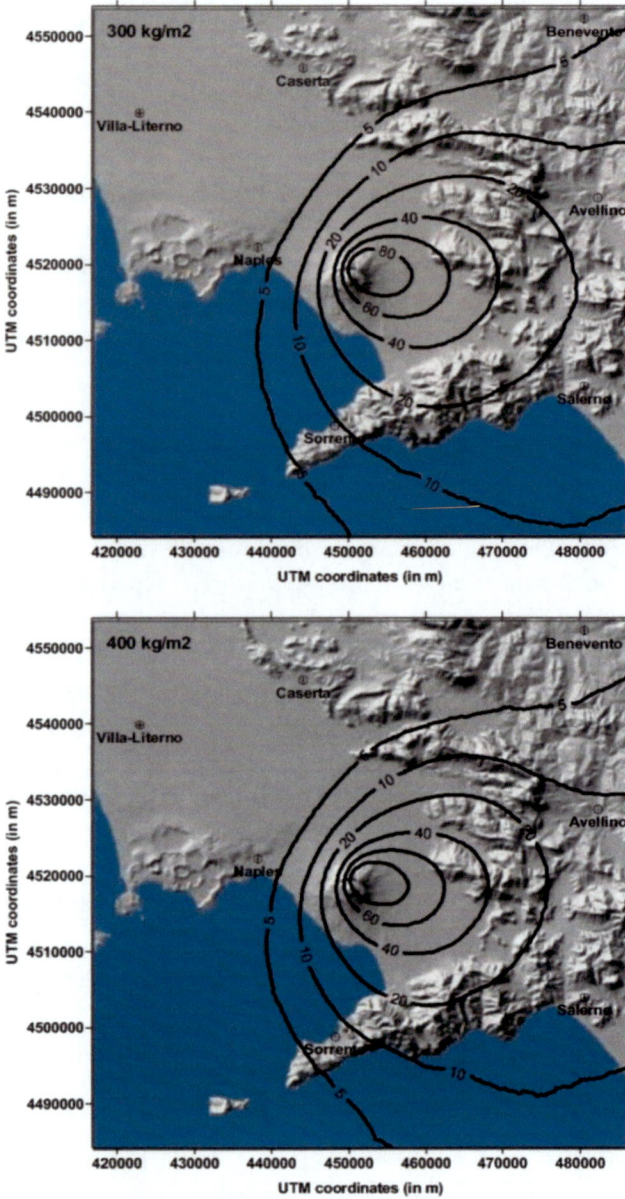

Figure 3. Ash loading probability maps for Plinian scenario. Probability is normalized to 100%. Top) 300 kg/m^2, bottom) 400 kg/m^2. (After Macedonio et al., 2008).

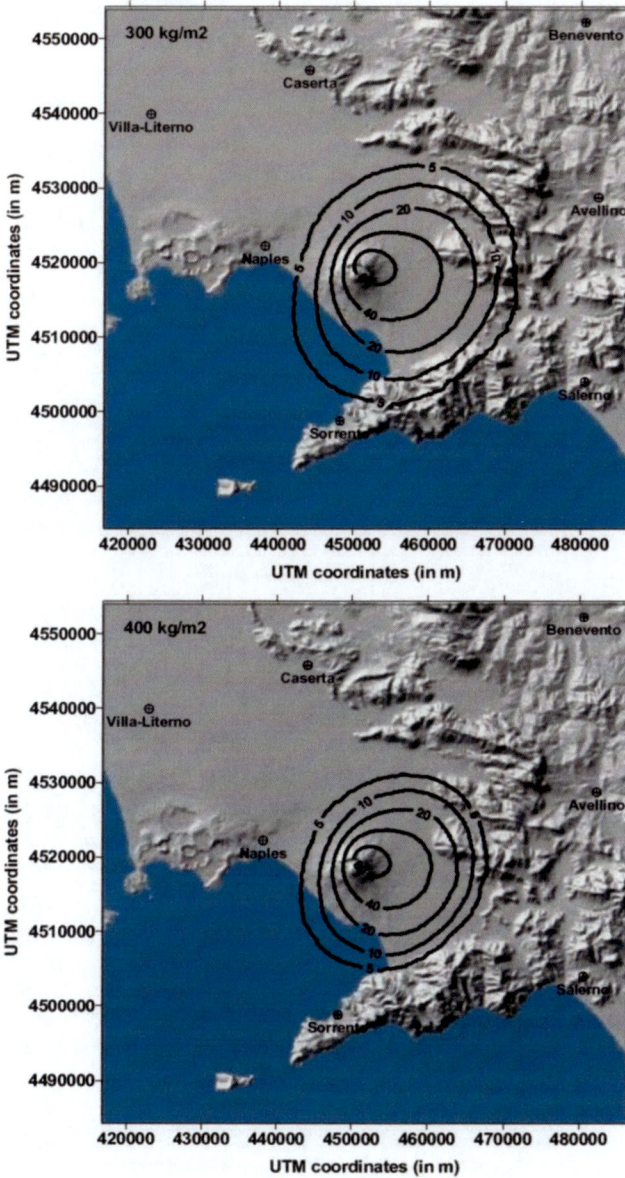

Figure 4. Ash loading probability maps for Sub-Plinian scenario. Probability is normalized to 100%. Top) 300 kg/m², bottom) 400 kg/m². (After Macedonio et al., 2008).

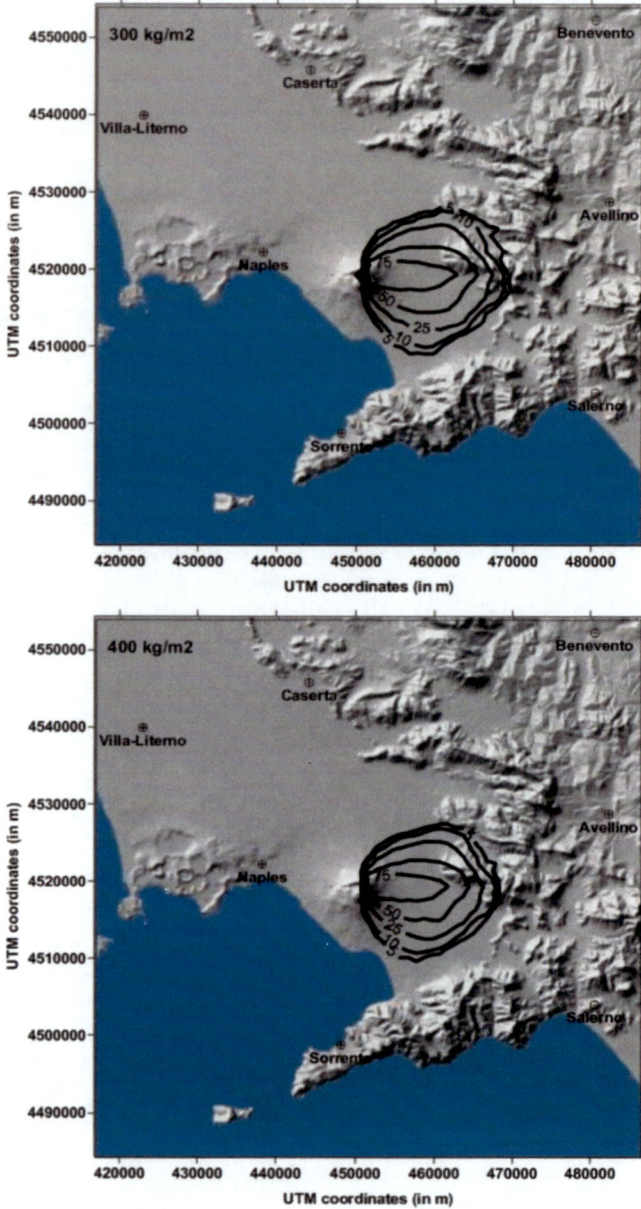

Figure 5. Ash loading probability maps for violent Strombolian scenario. Probability is normalized to 100%. Top) 300 kg/m^2, bottom) 400 kg/m^2. (After Macedonio et al., 2008).

A more recent work by Folch and Sulpizio (2010) presents the first tephra dispersal hazard assessment based on numerical modeling. Authors developed hazard and isochron maps for distal ash fallout for a scenario obtained combining the source parameters of different eruptive scenarios at Vesuvius, The scenario was defined on the base of the Maximum Expected Event (MEE, Barberi et al., 1990) characterized by an eruption volume that can be comparable to those erupted by the AD 472 and AD 1631 eruptions (Cioni et al., 2003a), classified as Sub-Plinian events.

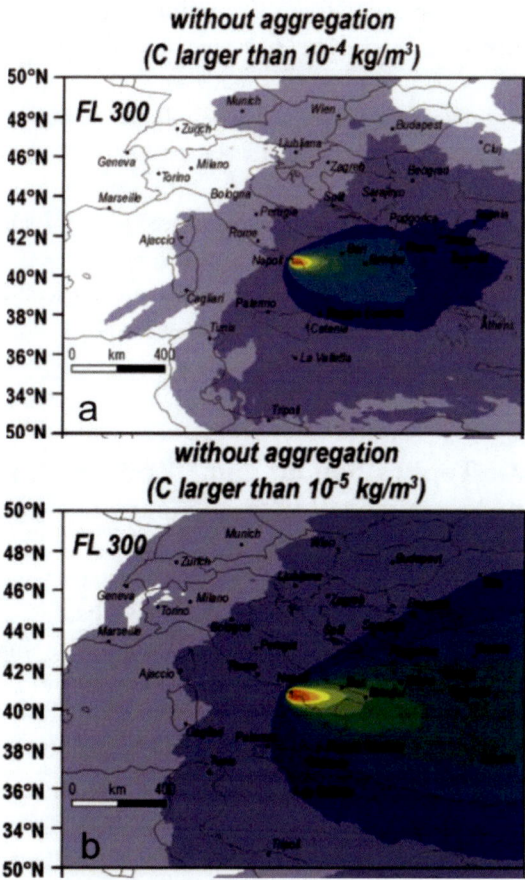

Figure 6. FL300 probability maps. The maps yield the probability (%) that the ash cloud concentration (C) exceeds 10^{-4} kg/m^3 (a) and 10^{-5} kg/m^3 (b). (Modified after Folch and Sulpizio, 2010).

Hazard maps at relevant flight levels were produced using super-computing facilities, spanning on a meteorological year statistically representative of the local meteorology during the last few decades (Folch and Sulpizio, 2010). The TTDM strategy relies on off-line coupling between a mesoscalar meteorological model (WRF, Michalakes et al., 2001) and the numerical TTDM FALL3D. Hazard maps were computed by calculating the probability that the ash cloud concentration overcomes a threshold value at a certain spatial point. The authors selected the critical threshold values of 10^{-4} and 10^{-5} g/m^3 and focused on the Flight Levels (FL) FL050 (5,000 feet) and FL300 (30,000 feet), representative of landing/take-off operations and jet cruise altitude, respectively. Figure 6 shows the resulting hazard maps at FL300. In case of occurrence of the considered Sub-Plinian scenario, a vast airspace region over the central Mediterranean area would be contaminated by ash, leading to potential disruptions of the busiest aerial corridors in the study area (Folch and Sulpizio, 2010). For the 10^{-4} g/m^3 concentration threshold, the low to moderate probability curves extend to most parts of southern Italy, and all of Macedonia, other than parts of Albania, Greece and Montenegro (Figure 6a). The lower concentration value (10^{-5} g/m^3) depicts a worse scenario, since the areas previously enclosed by the >10% probability curve, now have a probability greater than 30% (Figure 6b).

3.3. Probabilistic Tephra Fallout and Dispersal Hazard Assessment for Concepción Volcano, Nicaragua

Concepción volcano (Nicaragua) is a highly hazardous volcano. The recent geological record reveals an intense explosive activity characterized by Strombolian to sub-Plinian events, basaltic-to-dacitic in composition (Borgia and van Wyk de Vries, 2003). Due to the prevailing winds, past explosive eruptions have generated tephra deposits in the western part of the island that reached the shore of the Nicaragua Lake. One of the most copious fallout deposits is associated with the 1977 Concepción eruption (Delgado-Granados et al., 2006). Recent low-magnitude eruptive events caused minor tephra deposition at the western part of the island. Tephra fallout from explosive activity at the Concepción volcano can thus impact local communities in the Island and reach the heavily populated west coast of the Nicaragua Lake. Moreover, since the city of Managua, hosting the international airport, lays only 90 km from the volcano, an eruption of medium magnitude could easily affect the operability of the Managua airport. But, despite the high potential impacts of an explosive event at Concepción volcano, there is no official hazard map for tephra fallout and dispersal available for communities and civil protection authorities. The only existing hazard assessment from tephra fallout (Delgado-Granados et al. 2006) was based on isopachs from a few events combined with a statistical study of the prevailing regional winds. Scaini et al. (2012) extended the work of Delgado-Granados et al. (2006) producing a complete tephra fallout and dispersal hazard assessment for three different eruptive scenarios defined for the Concepción volcano based on the past historical record.

The tephra dispersal hazard assessment was performed at a local scale. This implied that a NWPM had to be run, although its usage would have involved large computing times and storage capacity. To circumvent these problems, Scaini et al. (2012) applied the Typical Meteorological Year (TMY) technique in order to constrain the representative meteorological conditions. A TMY consists of 12 representative months selected from individual years of a time period and collated in a meteorological database (e.g. Finkelstein and Schafer, 1971). The TMY method is used in many fields (climatology, solar power energy, etc.) but had never been used before in volcanology. In this work, the authors adapted the methodology for defining a TMY in terms of wind speed and direction, and identified the TMY for a 30-year period at three vertical intervals that correspond to the mean column height of the three eruptive scenarios considered.

The eruptive scenarios defined are: Low, Medium and High Magnitude Scenario (LMS, MMS and HMS, respectively), based on reference eruptions identified in the historical record. These scenarios correspond to the most significant eruptive styles expected for the Concepción volcano. For each scenario, each ESPs was defined by Gaussian probability distributions centered in the value assumed for the respective reference eruption (see Scaini et al., 2012 for details). This approach, introduced by Bonadonna (2006), allows accounting for uncertainties related with the scenario definition.

In order to initialize the TTDM model, ESPs must be sampled within their distribution (Bonadonna, 2006). However, extracting a representative subset of ESPs requires a high number of samples, that is, running many numerical simulations, which increases the computational cost of the hazard assessment. Scaini et al. (2012) introduce the *stratified sampling* technique (Rao and Krishnaiah, 1994), a statistic technique that can be applied when previous knowledge about the shape of a population of values is known or assumed, and consists of a random sampling that constrains the number of members within pre-defined bins.

The advantage, with respect to a purely random sampling, is that a stratified sampling gives a similar sampling accuracy with, at least, one order of magnitude less members, reducing the overall computational cost. TTDM simulations were then initialized with sets of sampled ESPs and 365*2 simulations were performed for each scenario.

Numerical simulations outputs were merged to produce tephra fallout and dispersal hazard maps for HMS. Results show that tephra load values are likely to reach considered critical thresholds (i.e. 1, 50 and 100 kg/m^2) over a wide area, and the collapse of buildings and structures are expected, especially in the island of Ometepe and in the Rivas province and, to a lesser extent, also in Granada and Managua. Figure 7 (top) shows an ash fallout hazard map for 1 kg/m^2.

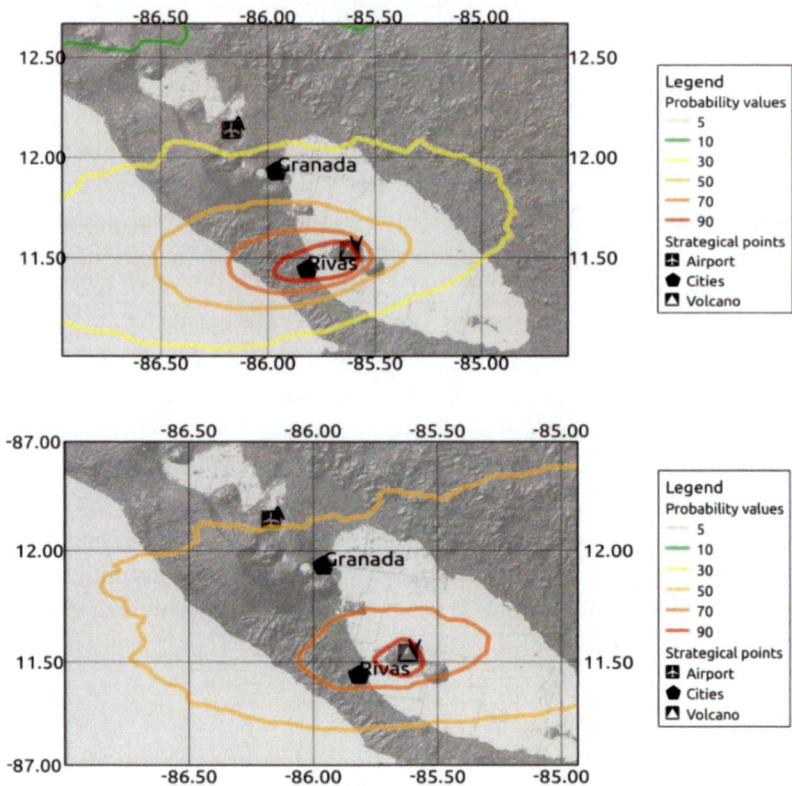

Figure 7. Top: ash fallout hazard map for a threshold value of 1 kg/m^2. Bottom: hazard map at FL050 for a critical ash concentration of 2 mg/m^3. (Modified after Scaini et al., 2012).

Hazard maps at FL050 (Figure 7, bottom) for a critical ash concentration of 2 mg/m^3 show that the central part of the country, including the capital, has a very high probability (around 50%) to have its airspace affected by the critical concentration used as no-fly limit during the 2010 aviation disruption in Europe. Given that HMS is likely to produce generalized disruption of ground transportation systems in the whole central part of Nicaragua, the contemporary closure of airspace may cause strong socio-economic impacts. In fact, it is worth mentioning that Managua International airport is the only international airport in Nicaragua and one of the busiest in Central America. Following Scaini et al. (2012) results, most Nicaraguan airspace has a high probability (up to 50%) of being affected by critical ash concentration in case of occurrence of HMS scenario. If such a critical situation lasted from days to weeks, it could cause serious disruption of the air transportation system.

Scaini et al. (2012) propose novel methods that allow producing a multi-scale hazard assessment for tephra fallout and dispersal, reducing the computational cost and accounting for uncertainties related to the definition of eruptive scenarios and results of their work should be taken into account for future planning at both local and regional scales.

3.4. Probabilistic Tephra Dispersal Hazard Assessment for Popocatepetl Volcano, Mexico

So far, the hazard assessment related to volcanic ash for Mexican volcanoes was carried out with semi-analytical models (Bonasia et al., 2011, in the case of the Colima volcano) or using a purely numerical model (Bonasia et al., 2012, in the case of the El Chichon volcano), but in both cases hazard maps were only related to ash load in proximal and intermediate zones. In particular, until recently there were no examples of modeling-based hazard mapping of tephra fallout and dispersal from the Popocatépetl volcano, which is considered one of the most active volcanoes in Mexico. The first long-range hazard assessment of volcanic ash dispersal for a Plinian eruptive scenario at Pocatépetl volcano was performed by Bonasia et al. (2014), with the main objective to increase public awareness regarding the enormous potential impact that volcanic ash could have on civil aviation in the region affected by the volcano.

The authors selected, as a possible Plinian eruptive scenario, the Ochre Pumice eruption (Arana-Salinas et al., 2010), which is the Popocatépetl's best-studied Holocene Plinian event. Their approach for the reconstruction of the probability maps follows the ERS (Bonadonna et al., 2005), which describes the probability of reaching a given tephra accumulation and dispersal based on the statistical distribution of both wind profiles and eruption source parameters.

Input eruptive parameters were constrained by means of the best-fit inversion method developed in Bonasia et al. (2010) that solves the inverse problem related to the reconstruction of tephra fallout deposits when only few sampling data are available.

Reconstructed parameters were associated to probability density functions and sampled using the *stratified sampling* technique (Rao and Krishnaiah, 1994; Scaini et al. 2012) in order to initialize the numerical code FALL3D (Costa et al., 2006; Folch et al., 2009). The meteorological inputs were randomly sampled within a 5-year period (2001-2006) from the NCEP/DOE Reanalysis 2 dataset (Kalnay et al., 1996).

As in Folch and Sulpizio (2010) and in Sulpizio et al. (2012), the probability maps of ash dispersal in the atmosphere were produced for two particular flight levels: FL050 and FL300, and for two ash concentration threshold values (2 and 0.2 mg/m^3) adopted in 2010 during the European crisis, as both are considered useful for civil aviation management purposes. Figures 8 and 9 show the hazard maps for FL050 and FL300 and the two concentration thresholds.

In every hazard map the highest probability contours involve Mexico City, Puebla and Veracruz, three international airports that exceed 27 million passengers per year. Moreover, aerial corridors over south-central Mexico would have a high probability of being disrupted or even interrupted with consequent socio-economic losses. The impact would not be limited to areas between the boundaries of Mexico but would include important international airports located in the south of United States, such as Miami, Dallas, and Houston, which together serve more than 150 million passengers per year.

The importance of addressing the hazard posed by a Plinian eruption at Popocatépetl volcano is increased by the fact that many other volcanoes in northern Central America have experienced Plinian eruptions during the Late Pleistocene-Holocene.

So, the renewal of Plinian activity at any of these volcanoes would probably have effects similar to those caused by Popocatépetl and the probability of ash dispersal maps computed for this volcano could serve as a starting point for future ash dispersal hazard evaluations at neighboring strato-volcanoes.

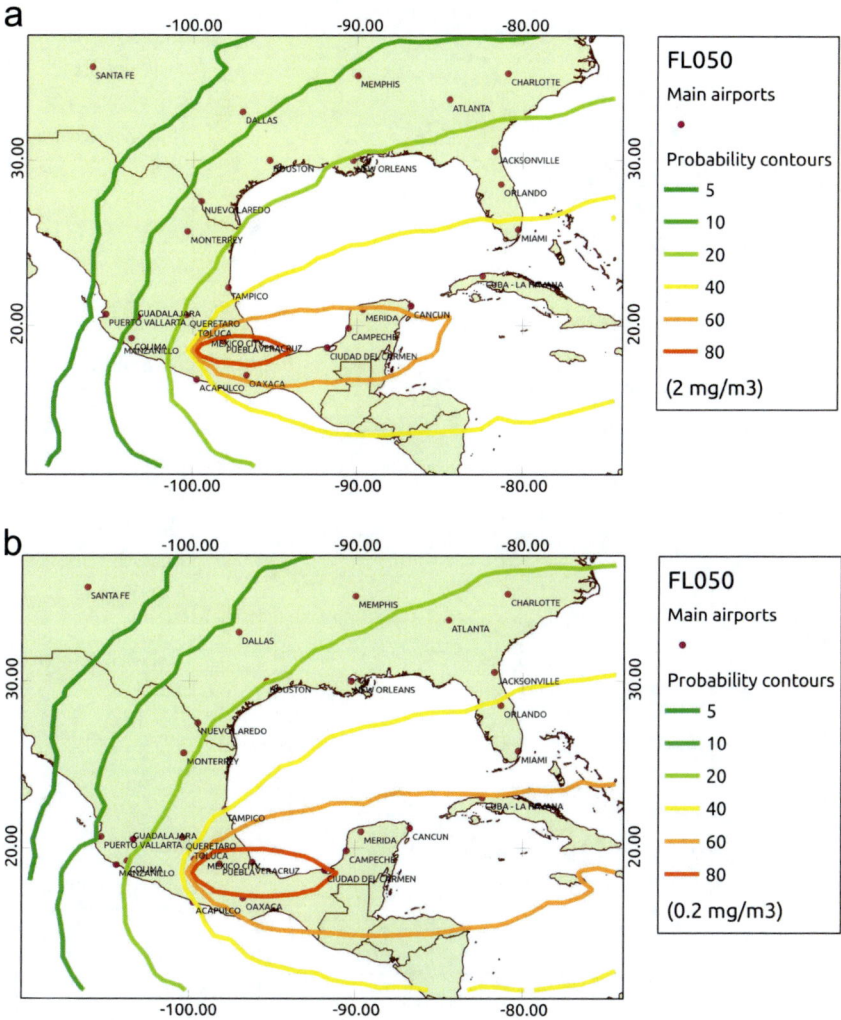

Figure 8. Probabilistic hazard maps at FL050, for two ash concentration thresholds: a) 2 mg/m³, b) 0.2 mg/m³. Contours give probability, in percent, of exceeding the given threshold value. (After Bonasia et al., 2014).

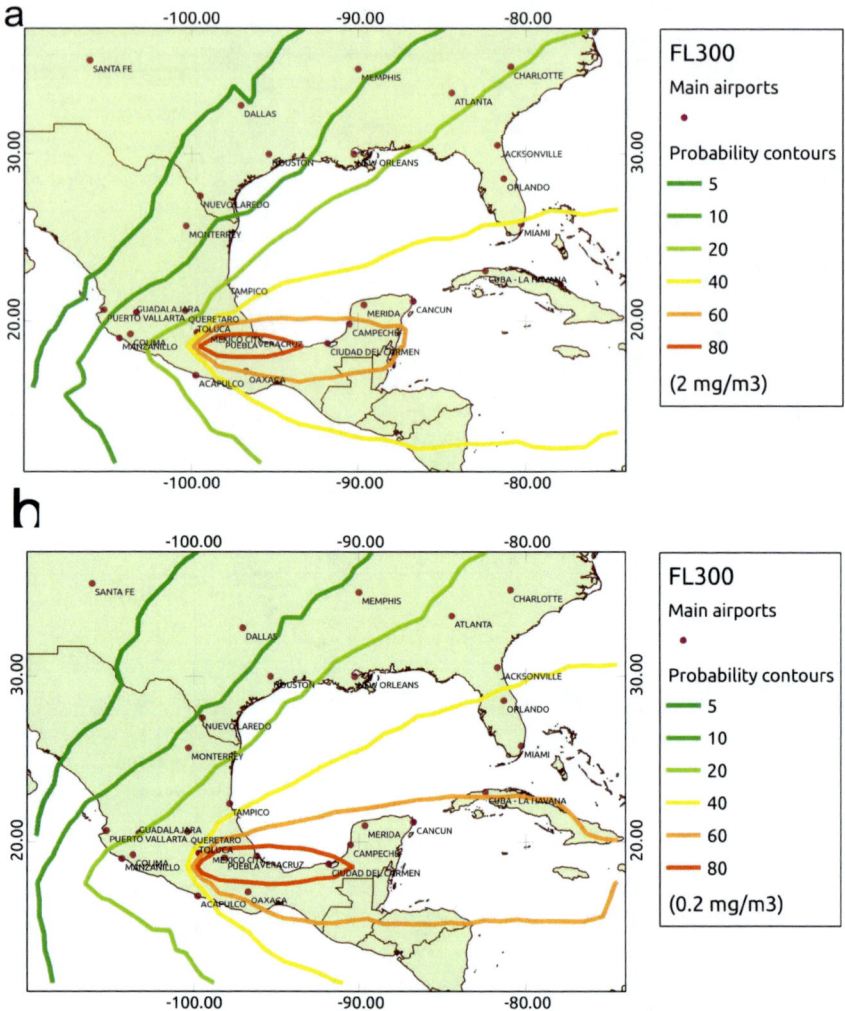

Figure 9. Probabilistic hazard maps at FL300, for two ash concentration thresholds: a) 2 mg/m³, b) 0.2 mg/m³. Contours give probability, in percent, of exceeding the given threshold value. (After Bonasia et al., 2014).

4. CONSIDERATIONS AND DEVELOPMENTS IN MODELING OF VOLCANIC ASH

After the 2010 aviation crisis, the attention posed to the volcanic ash dispersal hazard dramatically increased. As a response, the scientific community developed new techniques both for short-term forecasting and long-term volcanic ash dispersal and hazard assessment (Bonadonna et al., 2011; 2013) and enhanced the monitoring network at active volcanic areas. In this chapter, we described the advances introduced in recent years in the long-term hazard assessment strategy.

The main constraints on the production of probabilistic hazard maps are the computational cost of numerical simulations and the difficulty of characterizing the long-term meteorological and volcanological inputs. Initially, ESPs were maintained as constant (Folch and Sulpizio, 2010) or, in case of probabilistic hazard assessment, randomly sampled within their probability distribution (Bonadonna et al., 2006). The application of the *stratified sampling* technique (Rao and Krishaiah, 1994) to hazard assessment (Scaini et al. 2012) contributed to introducing the random sampling of ESPs to the tephra dispersal hazard assessment, and allowed reducing the number of simulations required for achieving a representative subset of input conditions. This technique was applied to recent tephra dispersal hazard assessments (Sulpizio et al., 2012; Bonasia et al., 2014) and can be applied to both meteorological and volcanological inputs.

In addition, the definition of representative meteorological conditions, based on vertical profiles applied for tephra fallout hazard assessment (Bonadonna et al., 2006; Macedonio et al., 2008; Costa et al., 2009), does not guarantee representation at the scale of the simulation. This choice is particularly critical for local-scale simulations, due to the high computational cost of running a mesoscalar model prior to TTDM. The first tephra dispersal hazard assessments (Folch and Sulpizio, 2010; Leadbetter and Hort, 2011) were based on the meteorological simulation performed for representative meteorological years selected *a priori*. But the choice of the representative conditions at a local scale has been improved by the introduction of the TMY that, although it has limitations (Scaini et al., 2012), supports the production of ash dispersal hazard assessment based on a statistically representative meteorological dataset.

Finally, the definition of the eruptive scenario is crucial and strongly affects the outputs of the hazard assessment. The definition of typical eruptive scenarios is based on the eruptive history of the volcano and the reconstruction of the eruptive history is done by means of field investigations. The complexity in the assessment of the proper eruptive scenario comes from the fact that explosive eruptions, besides being characterized by a wide range of hazardous phenomena, are also the result of different physical and geological processes that act over different time-scales. The parameterization of the physical processes that characterize the eruptive history of a volcano is affected by various uncertainties that mostly depend on the error associated with field measurements. As stated in Bonadonna et al. (2011), this problem can be overcome by improving the field investigation techniques. However, the uncertainty due to the random behaviour of the natural system can be solved only with the help of a probabilistic approach. Thus, a probabilistic analysis of eruptive records, when available, is an important aspect of the assessment of eruptive scenarios. The first step in the study of eruptive scenarios is to characterize eruptions by their magnitudes. Eruptive series are usually studied using conventional statistics, time-dependent series, or sequences including extreme events, which require special methods of analysis, such as the extreme value theory (De la Cruz-Reyna, 1993). Based on eruptive scenarios, probability distributions are defined for the ESPs, which allow accounting for the uncertainties related to the definition of scenarios. In addition, given that it is not always possible to associate a probability of occurrence to a given scenario, resulting hazard maps show, in most cases, the probability of having critical fallout/concentration values conditioned by the occurrence of the scenario considered. Thus, different hazard maps for different areas cannot be merged without accounting for this aspect (Biass et al., 2014).

Recent eruptions of the Eyjafjallajökull and Cordón de Caulle in 2010 and 2011, demonstrated the need for response strategies in order to lower socio-economic impacts of aviation disruptions, especially for long-lasting events. During the 2010 Eyjafjallajökull eruption, the precautionary "zero-ash tolerance" criterion for flight banning showed many limitations and led to the introduction of quantitative thresholds for allowing aviation operations (still under definition). Thus, the 2010 aviation disruption triggered the demand for accurate quantitative forecasts and related products. In addition, new regulations, such as the SRA, pointed out the importance of both short and long-term TTDM outputs for aviation purposes. The two IUGG-WMO workshops on ash dispersal forecast and civil aviation were organized to discuss the needs of the scientific community and aviation stakeholders involved in supporting aviation during volcanic eruptions. Results pointed out the need for improving and validating TTDM models in order to provide more reliable results to the aviation community (Bonadonna et al., 2011; 2013). In addition, uncertainties related to TTDM inputs and outputs should be characterized (Bonadonna et al., 2011; 2013). This can be achieved if model developers, meteorologists, volcanologists and stakeholders work to develop new strategies for the volcanic ash dispersal forecast. Moreover, we consider that more efficient communication between different communities is needed. Members of the science community have the responsibility of seeking research solutions that have reasonably low uncertainties and products that may be easily understood and used by the organisms prepared at civil protection purposes, such as for example, the Volcanic Ash Task Force, Civil Aviation Organization and Volcanic Ash Advisory Centres.

CONCLUSION

Hazard maps are of fundamental importance for increasing preparedness both in "peace time" and during unrest, before forecasts are available. The case studies presented here show that renewal of eruptive activity at the considered active volcanoes, would affect not only areas restricted to the vicinity of volcanoes, but also reach far beyond including neighboring countries with consequent great social and economic losses. The application of numerical TTDM for long-range hazard assessment is thus crucial in order to produce ash fallout and dispersal hazard maps at the most active volcanoes in the world. In addition, hazard maps serve as the basis for the first examples of vulnerability assessment and subsequent impact/risk analyses performed mostly for tephra fallout (Spence et al., 2008; Zuccaro et al., 2008; Biass et al., 2012; Jenkins et al., 2014) and recently introduced for ash dispersal (Scaini et al., 2014).

The success of the application of TTDM models for the computation of probabilistic hazard maps for ash fallout and dispersal can be achieved only with the establishment and strengthening of international collaborations between research and institutions that have an operational mandate, in order to effectively minimize social and economic impact of such eruptions. Research is of fundamental importance for developing new methodologies and techniques, but it has to be supported by institutions that have to agree on mutual expectations and requirements before a volcanic crisis.

REFERENCES

Barberi, F; Macedonio, G; Pareschi, MT; Santacroce, R. *Nature*, 1990, 344, 142-144.

Barberi, F; Ghigliotti, M; Macedonio, G; Orellana, H; Pareschi, MT; Rosi, M. *J Volcanol Geotherm Res*, 1992, 49, 53-68.

Biass, S; Frischknecht, C; Bonadonna, C. *Nat Hazards*, 2012, 64, 615-639.

Biass, S; Scaini, C; Bonadonna, C; Folch, A; Smith, K; Höskuldsson, A. *Nat Hazards Earth Syst Sci*, 2014 accepted.

Blong, RJ. *Volcanic Hazards*; Academic press, Sydney, 1984, 424.

Bolic, T; Sivcev, Z. Transportation Research Board Annual Meeting, January, 2011, Washington D. C. USA.

Bonadonna, C; Connor, CB; Houghton, BF; Connor, L; Byrne, M; Laing, A; Hincks, T. *J Geophys Res*, 2005, 110:B03203. doi:10.1029/2003JB002896.

Bonadonna, C; Folch, A; Loughlin, S; Puempel, H. *Bull Volcanol*, 2012, 74, 1-10.

Bonadonna, C; Webley, P; Hort, M; Folch, A; Loughlin, S; Puempel, H. Available at: www.unige.ch/sciences/terre/mineral/CERG/Workshop2/results-2/2nd-IUGG-WMO-WORKSHOP-CONS-DOC2.pdf. May 2014.

Bonasia R; Macedonio G; Costa A; Mele D; Sulpizio R. *J Volcanol Geotherm Res*, 2010, Doi: 10.1016/j.jvolgeores.2009.11.009.

Bonasia, R; Capra, L; Costa, A; Macedonio, G; Saucedo, R. *J Volcanol Geotherm Res*, 2011, 203, 12-22.

Bonasia, R; Costa, A; Folch, A; Macedonio, G; Capra, L. *J Volcanol Geotherm Res*, 2012, 231, 39-49.

Bonasia, R; Scaini, C; Capra, L; Nathenson, M; Siebe, C; Arana-Salinas, L; Folch, A. *Bull Volcanol*, 2013, 76, 789.

Borgia, A; Van Wyk de Vries, B. *Bull Volcanol*, 2003, 65, 248-266.

Casadevall, TJ. FAA *Aviat Saf J*, 1993, 2, 1-11.

Casadevall, TJ. in: discussion and recommendation from the workshop on impacts of volcanic ash on airports facilities Seattle Washington, 1993, 59.

Cioni, R; Levi, S; Sulpizio, R. *J Geophys Res*, 2003, 108, 2063.

Costa, A; Macedonio, G; Folch, A. *Earth Planet Sci Lett*, 2006, 241, 634-647.

Costa, A; Folch, A; Macedonio, G; Giaccio, B; Isaia, R; Smith, VC. *Geophys Res Lett*, 2012, 39, B09201.

De la Cruz-Reyna, S. *J Volcanol Geotherm Res*, 1993, 55, 51–68.

Delgado-Granados, H; Navarro, MT; Farraz, I; Alatorre Ibargüengoitia, MA; Hurst, AW. Fourth Conference Cities on Volcanoes 23-27 January 2006 Quito, Ecuador.

Favalli, M; Pareschi, MT; Zanchetta, G. *Bull Volcanol*, 2006, 6, 349-362.

Foch, A; Cavazzoni, C; Costa, A; Macedonio, G. *J Volcanol Geotherm Res*, 2008a, 177, 767-777.

Finkelstein, JM; Schafer, RE. *Biometrika*, 1971, 58, 641-645.

Folch, A; Jorba, O; Viramonte, J. *Nat Haz Earth Sys Scie*, 2008b, 8, 927-940.

Folch, A; Costa, A; Macedonia, G. Comput Geosci 2009, doi:10.1016/ j.cageo.2008.08.008.

Folch, A; Sulpizio, R. *Bull Volcanol*, 2010, 72, 1039-1059.

Guffanti, M; Mayberry, GC; Casadevall, TJ; Wunderman, R. *Nat Hazards*, 2009, 51(2),287–302.

Horwell, CJ; Baxter, PJ. *Bull Volcanol*, 2006, 69(1), 1–24.

Hurst, AW; Smith, W. *J Volcanol Geotherm Res*, 2004, 138, 393-403.

Hurst, AW; Turner, R. *J Geol Geophys*, 1999, 42, 615-622.

ICAO 2012 www.icao.int/publications/Pages/doc-series.aspx.

Jenkins, SF; Spence, RJS; Fonseca, JFBD; Solidum, RU; Wilson, TM. *J Volcanol Geotherm Res*, 2014, 276, 105-120.

Kalnay, E; Kanamitsu, M; Kister, R; Collins, W; Deaven, D; Gandin, L; Iredell, M; Saha, S; Woollen. J; Zhu,Y; Chelliah, M; Ebisuzaki, M; Higgins, W; Janowiak, J; Mo, K; Ropelewski, C; Wang, J; Leetmaa, A; Reynolds, R; Jenne, R; Joseph, D. *Bull Am Met Soc*, 1996, 77, 437-470.

Macedonio, G; Pareschi, MT; Santacroce, R. *J Geophys Res*, 1988, 93 (B12), 14817-14827.

Macedonio, G;Costa, A.: Longo, A. *Comput Ceosci*, 2005, 31, 837-845.

Macedonio, G; Costa, A; Folch, A. *J Volcanol Geotherm Res*, 2008, 178 (3), 366-377.

Mastin, lG; Guffanti, M; Sevranckx, R; Webley, P; Barsotti, S; Dean, K; Durant, A; Ewert, J.W; Neri, A; Rose, W.I; Schneider, D; Siebert, L; Stunder, B; Swanson, G; Tupper, A; Volentik, A; Waythomas, CF. *J Volcanol Geotherm Res*, 2009 186,10-21.

Michalakes, J; Chen, S; Dudhia, J; Hart, L; Klemp, J; Middlecoff, J; Skamarock, W. Developments in Terracomputing: Proceedings of the Ninth ECMWF Workshop on the Use of High Performance Computing in Meteorology, 2001, *World Scie Singapore*, 267-276.

Miller, TP; Casadevall, TJ. Encyclopedia of volcanoes In: Sigurdsson, H; Houghton, B; McNutt, S; Rymes, H; Stix, J. (eds) *Academic San Diego*, 2000, 915-930.

Nairn, IA; Shane, PR; Cole, JW; Leonard, GJ; Self, S; Pearson, N. *J Volcanol Geotherm Res*, 2004, 131(3-4), 265-294.

Papp, KR; Dean, KG; Dejn, J. *J Volcanol Geotherm Res*, 2005, 148, 295-314.

Paton, D; Millar, M; Johnston, D. *Nat Hazards*, 2001, 24, 157-179.

Pfeiffer, T; Costa, A; Macedonio, G. *J Volcanol Geotherm Res*, 2005, 140, 237-294.

Rao, CR; Krishnaiah, PR., In: Rao, CR., Krishnaiah, PR. (Eds.), Elsevier/North-Holland, Amsterdam, 125–145, 1994.

Scaini, C; Folch, A; Navarro, M. *J Volcanol Geotherm Res*, 2012, 219/220, 41-51.

Scaini, C; Biass, S; Galderisi, A; Bonadonna, C; Folch, A; Smith, K; Höskuldsson, A. *Nat Hazards Earth Sys Sci*, 2014 accepted.

Scollo, S; Prestifilippo, M; Spata, G; D'Agostino, M; Coltelli, M. *Nat Hazards Earth Sys Sci*, 2009, 9, 1573-1585.

Searcy, C; Dean, K; Stringer, W. *J Volcanol Geotherm Res*, 1998, 80, 1-16.

Spence, RJS; Kelman, I; Baxter, PJ; Zuccaro, G; Petrazzuoli, S. *Nat Hazards Earth Sys Sci*, 2005, 5,477–494.

Spence, R; Komorowski, JC; Saito, K; Brown, A; Pomonis, A; Toys, G; Baxter, P. *J Volcanol Geotherm Res*, 2008, 178, 516-528.

Stewart, C; Johnston, DM; Leonard, GS; Horwell, CJ; Thordarson, T; Cronin, SJ. *J Volcanol Geotherm Res*, 2006, 158, 296-306.

Stewart, C; Pizzolon, L; Wilson, T; Leonard, G; Denwar, D; Johnston, D; Cronin, S. *Integr Enviro Assess Manag*, 2009, 5, 713-716.

Sulpizio, R; Folch, A; Costa, A; Scaini, C; Dellino, P. *Bull Volcanol*, 2012, doi:10.1007/s00445-012-0656-3.

Suzuki, T. Volcanism: physics and tectonics In: Shimozuru, D; Yokoyama, I. (eds) *Terrapub Tokyo*, 1983, 95-113.

Wardman, JB; Wilson, TM; Bodger, PS; Cole, JW; Stewart, C. *Bull Volcanol*, 2012, doi:10.1007/s00445-012-0664-3.

Wilson, T; Cole, J; Cronin, S; Stewart, C; Johnston, D. *Nat Hazards*, 2011, 57,185–212, doi: 10.1007/s11069-010-9604-8.

Zuccaro, G; Cacace, F; Spence, R; Baxter, P. *J Volcanol Geotherm Res*, 2008, 178, 416-453.

Zuccaro, G; Leone, MF; Del Cogliano, D; Sgroi, A. *J Volcanol Geotherm Res*, 2013, 266, 1-15.

INDEX